PRAISE FOR
INSIDE OF A DOG

" . . . causes one's dog-loving heart to flutter with astonishment and gratitude."

—*The New York Times*

"A thoughtful take on the interior life of the dog . . . long on insight and short on jargon."

—*The Washington Post*

"An intelligent, fully dimensional portrait of the way dogs experience the world."

—*The Star-Ledger*

"An essential read for pet owners and students of animal behavior."

—*Library Journal*

"Nearly flawless."

—*The Bark*

"*Inside of a Dog* is a most welcomed authoritative, personal, and witty book about what it is like to be a dog. Alexandra Horowitz has spent a lot of time studying our best friends and shares her knowledge in a readable volume that also serves as a corrective to the many myths that circulate about just who our canine companions are. I hope this book enjoys the wide readership it deserves."

—Marc Bekoff, author of *The Emotional Lives of Animals* and *Wild Justice: The Moral Lives of Animals* (with Jessica Pierce)

"Discover why your dog is so sensitive to your emotions, gaze, and body language. Dogs live in a world of ever-changing intricate detail of smell. Read this captivating book and enter the sensory world of your dog."

—Temple Grandin, author of *Animals in Translation* and *Animals Make Us Human*

"This fascinating and elegant book will open your eyes, ears, and nose to what your dog thinks and understands. Part science, part personal observation, part love letter to a rescued dog named Pumpernickel—and all pure fun."

—Karen Pryor, author of *Reaching the Animal Mind*

INSIDE OF A DOG

WHAT DOGS SEE, SMELL, AND KNOW

ALEXANDRA HOROWITZ

SCRIBNER

New York   London   Toronto   Sydney

SCRIBNER
A Division of Simon & Schuster, Inc.
1230 Avenue of the Americas
New York, NY 10020

First Scribner trade paperback edition October 2010

SCRIBNER and design are registered trademarks of The Gale Group, Inc.,
used under license by Simon & Schuster, Inc., the publisher of this work.

For information about special discounts for bulk purchases,
please contact Simon & Schuster Special Sales:
1-866-506-1949 or business@simonandschuster.com.

The Simon & Schuster Speakers Bureau can bring authors to your live event.
For more information or to book an event contact the Simon & Schuster Speakers Bureau
at 1-866-248-3049 or visit our website at www.simonspeakers.com.

Manufactured in the United States of America

26 27 28 29 30

Library of Congress Control Number: 2008045842

ISBN 978-1-4165-8340-0
ISBN 978-1-4165-8343-1 (pbk)
ISBN 978-1-4165-8827-6 (ebook)

To the dogs

Contents

Contents

CONTENTS

Noble Mind

Inside of a Dog

You Had Me at Hello

The Importance of Mornings

CONTENTS

Outside of a dog, a book is man's best friend.
Inside of a dog, it's too dark to read.
— ATTRIBUTED TO GROUCHO MARX

INSIDE OF A DOG

Prelude

First you see the head. Over the crest of the hill appears a muzzle, drooling. It is as yet not visibly attached to anything. A limb jangles into view, followed in unhasty succession by a second, third, and fourth, bearing a hundred and forty pounds of body between them. The wolfhound, three feet at his shoulder and five feet to his tail, spies the long-haired Chihuahua, half a dog high, hidden in the grasses between her owner's feet. The Chihuahua is six pounds, each of them trembling. With one languorous leap, his ears perked high, the wolfhound arrives in front of the Chihuahua. The Chihuahua looks demurely away; the wolfhound bends down to Chihuahua level and nips her side. The Chihuahua looks back at the hound, who raises his rear end up in the air, tail held high, in preparation to attack. Instead of fleeing from this apparent danger, the Chihuahua matches his pose and leaps onto the wolfhound's face, embracing his nose with her tiny paws. They begin to play.

For five minutes these dogs tumble, grab, bite, and lunge at each other. The wolfhound throws himself onto his side and the little dog responds with attacks to his face, belly, and paws. A swipe by the hound sends the Chihuahua scurrying backward, and she timidly sidesteps out of his reach. The hound barks, jumps up, and arrives back on his feet with a thud. At this, the Chihuahua races toward one of those feet and bites it, hard. They are in mid-embrace—the hound with his mouth surrounding the body of the Chihuahua, the Chihuahua kicking back at the hound's face—when an owner snaps a leash on the hound's collar and pulls him upright and away. The Chihuahua rights herself, looks after them, barks once, and trots back to her owner.

These dogs are so incommensurable with each other that they may as well be different species. The ease of play between them always puzzled me. The wolfhound bit, mouthed, and charged at the Chihuahua; yet the little dog responded not with fright but in kind. What explains their ability to play together? Why doesn't the hound see the Chihuahua as prey? Why doesn't the Chihuahua see the wolfhound as predator? The answer turns out to have nothing to do with the Chihuahua's delusion of canine grandeur or the hound's lack of predatory drive. Neither is it simply hardwired instinct taking over.

There are two ways to learn how play works—and what playing dogs are thinking, perceiving, and saying: be born as a dog, or spend a lot of time carefully observing dogs. The former was unavailable to me. Come along as I describe what I've learned by watching.

I am a dog person.

My home has always had a dog in it. My affinity for dogs

began with our family dog, Aster, with his blue eyes, lopped tail, and nighttime neighborhood ramblings that often left me up late, wearing pajamas and worry, waiting for his midnight return. I long mourned the death of Heidi, a springer spaniel who ran with excitement—my childhood imagination had her tongue trailing out of the side of her mouth and her long ears blown back with the happy vigor of her run—right under a car's tires on the state highway near our home. As a college student, I gazed with admiration and affection at an adopted chow mix Beckett as she stoically watched me leave for the day.

And now at my feet lies the warm, curly, panting form of Pumpernickel—*Pump*—a mutt who has lived with me for all of her sixteen years and through all of my adulthood. I have begun every one of my days in five states, five years of graduate school, and four jobs with her tail-thumping greeting when she hears me stir in the morning. As anyone who considers himself a dog person will recognize, I cannot imagine my life without this dog.

I am a dog person, a lover of dogs. I am also a scientist.

I study animal behavior. Professionally, I am wary of anthropomorphizing animals, attributing to them the feelings, thoughts, and desires that we use to describe ourselves. In learning how to study the behavior of animals, I was taught and adhered to the scientist's code for describing actions: be objective; do not explain a behavior by appeal to a mental process when explanation by simpler processes will do; a phenomenon that is not publicly observable and confirmable is not the stuff of science. These days, as a professor of animal behavior, comparative cognition, and psychology, I teach from masterful texts that deal in quantifiable fact. They describe everything from hormonal and genetic explanations for the social behavior of animals, to conditioned responses, fixed action patterns, and optimal foraging rates, in the same steady, objective tone.

And yet.

Most of the questions my students have about animals remain quietly unanswered in these texts. At conferences where I have presented my research, other academics inevitably direct the postlecture conversations to their own experiences with their pets. And I still have the same questions I'd always had about my own dog—and no sudden rush of answers. Science, as practiced and reified in texts, rarely addresses our experiences of living with and attempting to understand the minds of our animals.

In my first years of graduate school, when I began studying the science of the mind, with a special interest in the minds of non-human animals, it never occurred to me to study dogs. Dogs seemed so familiar, so understood. There is nothing to be learned from dogs, colleagues claimed: dogs are simple, happy creatures whom we need to train and feed and love, and that is all there is to them. There is no *data* in dogs. That was the conventional wisdom among scientists. My dissertation advisor studied, respectably, baboons: primates are the animals of choice in the field of animal cognition. The assumption is that the likeliest place to find skills and cognition approaching our own is in our primate brethren. That was, and remains, the prevailing view of behavioral scientists. Worse still, dog owners seemed to have already covered the territory of theorizing about the dog mind, and their theories were generated from anecdotes and misapplied anthropomorphisms. The very notion of the mind of a dog was tainted.

And yet.

I spent many recreational hours during my years of graduate school in California in the local dog parks and beaches with Pumpernickel. At the time I was in training as an ethologist, a scientist of animal behavior. I joined two research groups observing highly social creatures: the white rhinoceros at the Wild

Animal Park in Escondido, and the bonobos (pygmy chimpanzees) at the Park and the San Diego Zoo. I learned the science of careful observations, data gathering, and statistical analysis. Over time, this way of looking began seeping into those recreational hours at the dog parks. Suddenly the dogs, with their fluent travel between their own social world and that of people, became entirely unfamiliar: I stopped seeing their behavior as simple and understood.

Where I once saw and smiled at play between Pumpernickel and the local bull terrier, I now saw a complex dance requiring mutual cooperation, split-second communications, and assessment of each other's abilities and desires. The slightest turn of a head or the point of a nose now seemed directed, meaningful. I saw dogs whose owners did not understand a single thing their dogs were doing; I saw dogs too clever for their playmates; I saw people misreading canine requests as confusion and delight as aggression. I began bringing a video camera with us and taping our outings at the parks. At home I watched the tapes of dogs playing with dogs, of people ball- and Frisbee-tossing to their dogs—tapes of chasing, fighting, petting, running, barking. With new sensitivity to the possible richness of social interactions in an entirely non-linguistic world, all of these once ordinary activities now seemed to me to be an untapped font of information. When I began watching the videos in extremely slow-motion playback, I saw behaviors I had never seen in years of living with dogs. Examined closely, simple play frolicking between two dogs became a dizzying series of synchronous behaviors, active role swapping, variations on communicative displays, flexible adaptation to others' attention, and rapid movement between highly diverse play acts.

What I was seeing were snapshots of the minds of the dogs, visible in the ways they communicated with each other and tried to communicate with the people around them—and, too, in the way they interpreted other dogs' and people's actions.

I never saw Pumpernickel—or any dog—the same way again. Far from being a killjoy on the delights of interacting with her, though, the spectacles of science gave me a rich new way to look at what she was doing: a new way to understand life as a dog.

Since those first hours of viewing, I have studied dogs at play: playing with other dogs and playing with people. At the time I was unwittingly part of a sea change taking place in science's attitude toward studying dogs. The transformation is not yet complete, but the landscape of dog research is already remarkably different than it was twenty years ago. Where once there was an inappreciable number of studies of dog cognition and behavior, there are now conferences on the dog, research groups devoted to studying the dog, experimental and ethological studies on the dog in the United States and abroad, and dog research results sprinkled through scientific journals. The scientists doing this work have seen what I have seen: the dog is a perfect entry into the study of non-human animals. Dogs have lived with human beings for thousands, maybe hundreds of thousands of years. Through the artificial selection of domestication, they have evolved to be sensitive to just those things that importantly make up our cognition, including, critically, attention to others.

In this book I introduce you to the science of the dog. Scientists working in laboratories and in the field, studying working dogs and companion dogs, have gathered an impressive amount of information on the biology of dogs—their sensory abilities, their behavior—and on the psychology of dogs—their cognition. Drawing from the accumulated results of hundreds of research programs, we can begin to create a picture of the dog from the inside—of the skill of his nose, what he hears, how his eyes turn to us, and the brain behind it all. The dog cognition work reviewed includes my own but extends far beyond it to summarize all the results from recent research. For some topics on which there is no reliable information yet on dogs, I incorporate studies on other animals that might help us understand a dog's life, too. (For those whose appetite for the original research articles is whetted by the accounts herein, full citations appear at the book's end.)

We do no disservice to dogs by stepping away from the leash and considering them scientifically. Their abilities and point of view merit special attention. And the result is magnificent: far from being distanced by science, we are brought closer to and can marvel at the true nature of the dog. Used rigorously but creatively, the process and results of science can shed new light on discussions that people have daily about what their dog knows, understands, or believes. Through my personal journey, learning to look systemically and scientifically at my own dog's behavior, I came to have a better understanding of, appreciation of, and relationship with her.

I've gotten inside of the dog, and have glimpsed the dog's point of view. You can do the same. If you have a dog in the room with you, what you see in that great, furry pile of dogness is about to change.

A PREFATORY NOTE ON THE DOG,
TRAINING, AND OWNERS

Calling a dog "the dog"

It is the nature of scientific study of non-human animals that a few individual animals who have been thoroughly poked, observed, trained, or dissected come to represent their entire species. Yet with humans we never let one person's behavior stand for all of our behavior. If one man fails to solve a Rubik's cube in an hour, we do not extrapolate from that that all men will so fail (unless that man had bested every other man alive). Here our sense of individuality is stronger than our sense of shared biology. When it comes to describing our potential physical and cognitive capacities, we are individuals first, and members of the human race second.

By contrast, with animals the order is reversed. Science considers animals as representatives of their species first, and as individuals second. We are accustomed to seeing a single animal or two kept in a zoo as representative of their species; to zoo management, they are even unwitting "ambassadors" of the species. Our view of the uniformity of species members is well exemplified in our comparison of their intelligence. To test the hypothesis, long popular, that having a bigger brain indicates greater intelligence, the brain volumes of chimpanzees, monkeys, and rats were compared with human brains. Sure enough, the chimp's brain is smaller than ours, the monkey's smaller than the chimp's, the rat's a mere cerebellum-sized node of the primates' brains. That much of the story is fairly well known. What is more surprising is that the brains used, for comparative purposes, were the brains of just two or three chimpanzees and monkeys. These couple of animals unlucky enough to lose their heads for science were henceforth considered perfectly representative monkeys and chimps. But we had no idea if they happened to be particularly big-brained monkeys, or abnormally small-brained chimps.*

Similarly, if a single animal or small group of animals fails in a psychological experiment, the species is tainted with the brush of failure. Although grouping animals by biological similarity is clearly useful shorthand, there is a strange result: we tend to speak of the species as though all members of the species were identical. We never make this slip with humans. If a dog, given the choice between a pile of twenty biscuits and a pile of ten biscuits, chooses the latter, the conclusion is often stated with the definite article: "the dog" cannot distinguish between large and small piles—not "a dog" cannot so distinguish.

So when I talk about *the dog,* I am talking implicitly about *those dogs studied to date.* The results of many well-performed experiments may eventually allow us to reasonably generalize to *all dogs,* period. But even then, the variations among individual dogs will be great: your dog may be an unusually good smeller, may never look you in the eye, may love his dog bed and hate to be touched. Not every behavior a dog does should be interpreted as telling, taken as something intrinsic or fantastic; sometimes they just *are,* just as we are. That said, what I offer herein is the known capacity of *the dog;* your results may be different.

---

*Of course, researchers soon found brains bigger than ours: the dolphin's brain is larger, as are the brains of physically larger creatures such as whales and elephants. The "big brain" myth has long been overturned. Those who are still interested in mapping brain to smarts now look at other, more sophisticated measures: the amount of convolution of the brain; the encephalization quotient, a ratio that includes both brain and body size in the calculation; the quantity of neocortex; or the gross number of neurons and synapses between neurons.

Training dogs

This is not a dog training book. Still, its contents might lead you
to be able to train your dog, inadvertently. This will catch us up
to dogs, who have already, without a tome on people, learned
how to train us without our realizing it.

The dog training literature and the dog cognition and
behavior literatures do not overlap greatly. Dog trainers do use
a few basic tenets from psychology and ethology—sometimes to
great effect, sometimes to disastrous end. Most training operates
on the principle of *associative learning*. Associations between
events are easily learned by all animals, including humans. Associative learning is what is behind "operant" conditioning paradigms, which provide a reward (a treat, attention, a toy, a pat)
after the occurrence of a desired behavior (a dog sitting down).
Through repeated application, one can *shape* a new, desired
behavior in a dog—be it lying down and rolling over, or, for the
ambitious, calmly Jet-Skiing behind a motorboat.

But often the tenets of training clash with the scientific study
of the dog. For instance, many trainers use the analogy of dog-
as-tame-wolf as informative in how we should see and treat
dogs. An analogy can only be as good as its source. In this case,
as we will see, scientists know a limited amount about natural
wolf behavior—and what we know often contradicts the conventional wisdom used to bolster those analogies.

In addition, training methods are not scientifically tested,
despite some trainers' assertions to the contrary. That is, no
training program has been evaluated by comparing the performance of an experimental group that gets training and a
control group whose life is identical except for the absence of the
training program. People who come to trainers often share two

unusual features: their dogs are less "obedient" than the average dog, and the owners are more motivated to change them than the average owner. It is very likely, given this combination of conditions and a few months, that the dog will behave differently after training, almost regardless of what the training is.

Training successes are exciting, but they do not prove that the training method is what led to the success. The success *could* be indicative of good training. But it could also be a happy accident. It could also be the result of more attention being paid to the dog over the course of the program. It could be the result of the dog's maturing over the course of the program. It could be the result of that bullying dog down the street moving away. In other words, the success could be the result of dozens of other co-occurring changes in the dog's life. We cannot distinguish these possibilities without rigorous scientific testing.

Most critically, training is usually tailored to the owner—to change the dog to fit the owner's conception of the role of the dog, and of what he wants the dog to do. This goal is quite different than our aim: looking to see what the dog actually does, and what he wants from and understands of you.

The dog and his owner

It is increasingly in vogue to speak not of pet ownership but pet guardianship, or pet companions. Clever writers talk of dogs' "humans," turning the ownership arrow back on ourselves. In this book I call dogs' families *owners* simply because this term describes the legal relationship we have with dogs: peculiarly, they are still considered property (and property of little compensatory value, besides breeding value, a lesson I hope no reader ever has to learn personally). I will celebrate the day when dogs are not property which we own. Until then, I use the word *owner* apolitically, for convenience and with no other motive. This motive guides me in my pronoun use, too: unless discussing a female dog, I usually call the dog "him," as this is our gender-neutral term. The reputedly more neutral "it" is not an option, for anyone who has known a dog.

Umwelt: From the Dog's Point of Nose

This morning I was awakened by Pump coming over to the bed and sniffing emphatically at me, millimeters away, her whiskers grazing my lips, to see if I was awake or alive or me. She punctuates her rousing with an exclamatory sneeze directly in my face. I open my eyes and she is gazing at me, smiling, panting a hello.

Go look at a dog. Go on, look—maybe at one lying near you right now, curled around his folded legs on a dog bed, or sprawled on his side on the tile floor, paws flitting through the pasture of a dream. Take a good look—and now forget everything you know about this or any dog.

This is admittedly a ridiculous exhortation: I don't really expect that you could easily forget even the name or favored food or unique profile of your dog, let alone everything about him. I think of the exercise as analogous to asking a newcomer to meditation to enter into satori, the highest state, on the first go: aim for it, and see how far you get. Science, aiming for objectivity, requires that one becomes aware of prior prejudices and personal perspective. What we'll find, in looking at dogs through a scientific lens, is that some of what we think we know about dogs is entirely borne out; other things that appear patently true are, on closer examination, more doubtful than we thought. And by looking at our dogs from another perspective—from the perspective of the dog—we can see new things that don't naturally occur to those of us encumbered with human brains. So the best way to begin understanding dogs is by forgetting what we think we know.

The first things to forget are anthropomorphisms. We see, talk about, and imagine dogs' behavior from a human-biased perspective, imposing our own emotions and thoughts on these furred creatures. *Of course,* we'll say, dogs love and desire; of course they dream and think; they also know and understand us, feel bored, get jealous, and get depressed. What could be a more natural explanation of a dog staring dolefully at you as you leave the house for the day than that he is depressed that you're going?

The answer is: an explanation based in what dogs actually have the capacity to feel, know, and understand. We use these words, these anthropomorphisms, to help us make sense of dogs' behavior. Naturally, we are intrinsically prejudiced toward human experiences, which leads us to understand animals' experiences only to the extent that they match our own. We remember stories that confirm our descriptions of animals and conveniently forget those that do not. And we do not hesitate to assert "facts" about apes or dogs or elephants or any animal without proper evidence. For many of us, our interaction with non-pet animals begins and ends with our staring at them at zoos or watching shows on cable TV. The amount of useful information we can get from this kind of eavesdropping is limited: such a passive encounter reveals even less than we get from glancing in a neighbor's window as we walk by.* At least the neighbor is of our own species.

Anthropomorphisms are not inherently odious. They are born of attempts to understand the world, not to subvert it. Our human ancestors would have regularly anthropomorphized in an attempt to explain and predict the behavior of other animals, including those they might want to eat or that might want to eat them. Imagine encountering a strange, bright-eyed jaguar at dusk in the forest, and looking squarely in its eyes looking squarely into yours. At that moment, a little meditation on what you might be thinking "if you were the jaguar" would probably be due—and would lead to your hightailing it away from the cat. Humans endured: the attribution was, if not true, at least true enough.

Typically, though, we are no longer in the position of needing to imagine the jaguar's desires in time to escape his clutches. Instead we are bringing animals inside and asking them to become members of our families. For that purpose, anthropomorphisms fail to help us incorporate those animals into our homes, and have the smoothest, fullest relationships with them. This is not to say that we're always wrong with our attributions: it might be true that our dog is sad, jealous, inquisitive,

---

*This was made most evident for me one day collecting data of the behavior of the white rhinoceros. At the Wild Animal Park it is the animals who roam (relatively) freely, and the visitors are restricted to trains that travel around the large enclosures. I was situated in the narrow patch of grass between the track and the fence, watching a typical day of rhino socializing. As the trains approached, the rhinos stopped what they were doing and moved quickly into a defensive huddle: standing with rumps together, heads radiating out in a rough sunburst. The animals are peaceful, but with poor vision they can be easily startled if they do not smell someone approaching, and they count on each other as lookouts. The train stopped, and everyone gaped at the rhinos who, it was announced by the guide, were "doing nothing." Eventually the driver moved on, and the rhinos resumed their ordinary behavior.

depressed—or desiring a peanut butter sandwich for lunch. But we are almost certainly not justified in claiming, say, *depression* from the evidence before us: the mournful eyes, the loud sigh. Our projections onto animals are often impoverished—or entirely off the mark. We might judge an animal to be happy when we see an upturn of the corners of his mouth; such a "smile," however, can be misleading. On dolphins, the smile is a fixed physiological feature, immutable like the creepily painted face of a clown. Among chimpanzees, a grin is a sign of fear or submission, the furthest thing from happiness. Similarly, a human might raise her eyebrows in surprise, but the eyebrow-raising capuchin monkey is not surprised. He is evincing neither skepticism nor alarm; instead he is signaling to nearby monkeys that he has friendly designs. By contrast, among baboons a raised brow can be a deliberate threat (lesson: be careful which monkey you raise your eyebrows toward). The onus is on us to find a way to confirm or refute these claims we make of animals.

It may seem a benign slip from sad eyes to depression, but anthropomorphisms often slide from benign to harmful. Some risk the welfare of the animals under consideration. If we're to put a dog on antidepressants based on our interpretation of his eyes, we had better be pretty sure of our interpretation. When we assume we know what is best for an animal, extrapolating from what is best for us or any person, we may inadvertently be acting at cross-purposes with our aims. For instance, in the last few years there has been considerable to-do made about improved welfare for animals raised for food, such as broiler chickens who have access to the outside, or have room to roam in their pens. Though the end result is the same for the chicken—it winds up as someone's dinner—there is a budding interest in the welfare of the animals before they are killed.

But do they want to range freely? Conventional wisdom

holds that no one, human or not, *likes* to be pressed up against others. Anecdotes seem to confirm this: given the choice of a subway car jammed with hot, stressed commuters, and one with only a handful of people, we choose the latter in a second (heeding the possibility, of course, that there's some other explanation—a particularly smelly person, or a glitch in air-conditioning—that explains this favorable distribution). But the natural behavior of chickens may indicate otherwise: chickens flock. They don't sally forth on their own.

Biologists devised a simple experiment to test the chickens' preferences of where to be: they picked up individual animals, relocated them randomly within their houses, and monitored what the chickens did next. What they found was that most chickens moved closer to other chickens, not farther away, even when there was open space available. Given the option of space to spread their wings . . . they choose the jammed subway car.

This is not to say that chickens thus *like* being smushed against other birds in a cage, or find it a perfectly agreeable life. It is inhumane to pen chickens so tightly they cannot move. But it is to say that assuming resemblance between chicken preferences and our preferences is not the way to insight about what the chicken actually does like. Not coincidentally, these broiler chickens are killed before they reach six weeks of age; domestic chicks are still being brooded by their mothers at that age. Deprived of the ability to run under her wings, the broiler chickens run closer to other chickens.

TAKE MY RAINCOAT. PLEASE.

Do our anthropomorphic tendencies ever miss so fabulously with dogs? Without a doubt they do. Take raincoats. There are some interesting assumptions involved in the creation and purchase of tiny, stylish, four-armed rain slickers for dogs. Let's put aside the question of whether dogs prefer a bright yellow slicker, a tartan pattern, or a raining-cats-and-dogs motif (*clearly* they prefer the cats and dogs). Many dog owners who dress their dogs in coats have the best intentions: they have noticed, perhaps, that their dog resists going outside when it rains. It seems reasonable to extrapolate from that observation to the conclusion that he *dislikes* the rain.

He dislikes the rain. What is meant by that? It is that he must dislike *getting the rain on his body,* the way many of us do. But is that a sound leap? In this case, there is plenty of seeming evidence from the dog himself. Is he excited and wagging when you get the raincoat out? That seems to support the leap . . . or, instead, the conclusion that he realizes that the appearance of the coat predicts a long-awaited walk. Does he flee from the coat? Curl his tail under his body and duck his head? Undermines the leap—though does not discredit it outright. Does he look bedraggled when wet? Does he shake the water off excitedly? Neither confirmatory nor disconfirming. The dog is being a little opaque.

Here the natural behavior of related, wild canines proves the most informative about what the dog might think about a raincoat. Both dogs and wolves have, clearly, their own coats permanently affixed. One coat is enough: when it rains, wolves may seek shelter, but they do not cover themselves with natural materials. That does not argue for the need for or interest in raincoats. And besides being a jacket, the raincoat is also one distinctive thing: a close, even pressing, covering of the back, chest, and sometimes the head. There *are* occasions when wolves get pressed upon the back or head: it is when they are being dominated by another wolf, or scolded by an older wolf or relative.

Dominants often pin subordinates down by the snout. This is called muzzle biting, and accounts, perhaps, for why muzzled dogs sometimes seem preternaturally subdued. And a dog who "stands over" another dog is being dominant. The subordinate dog in that arrangement would feel the pressure of the dominant animal on his body. The raincoat might well reproduce that feeling. So the principal experience of wearing a coat is not the experience of feeling protected from wetness; rather, the coat produces the discomfiting feeling that someone higher ranking than you is nearby.

This interpretation is borne out by most dogs' behavior when getting put into a raincoat: they may freeze in place as they are "dominated." You might see the same behavior when a dog resisting a bath suddenly stops struggling when he gets fully sodden or covered with a heavy, wet towel. The be-jacketed dog may cooperate in going out, but not because he has shown he likes the coat; it is because he has been subdued.* And he will wind up being less wet, but it is we who care about the planning for that, not the dog. The way around this kind of misstep is to replace our anthropomorphizing instinct with a behavior-reading instinct. In most cases, this is simple: we must ask the dog what he wants. You need only know how to translate his answer.

---

*This is similar to what was discovered by midcentury behaviorist researchers who exposed laboratory dogs to an electric shock from which they couldn't escape. Later, put in a chamber from which there was a visible escape route, and shocked again, these dogs showed *learned helplessness:* they did not try to avoid the shock by escaping. Instead, they froze in place, seemingly resigned to their fate. The researchers had essentially trained the dogs to be submissive and accept their lack of control of the situation. (They later forced the dogs to unlearn the response and end the shock.) Happily, the days of experiments wherein we shock dogs to learn about their responses are over.

A TICK'S VIEW OF THE WORLD

Here is our first tool to getting that answer: imagining the point of view of the dog. The scientific study of animals was changed by a German biologist of the early twentieth century named Jakob von Uexküll. What he proposed was revolutionary: anyone who wants to understand the life of an animal must begin by considering what he called their *umwelt* (OOM-velt): their subjective or "self-world." Umwelt captures what life is like *as* the animal. Consider, for instance, the lowly deer tick. Those of you who have spent long minutes hesitatingly petting the body of a dog for the telltale pinhead that indicates a tick swollen with blood may have already considered the tick. And you probably consider the tick as a pest, period. Barely even an animal. Von Uexküll considered, instead, what it might be like from the tick's point of view.

A little background: ticks are parasites. Members of Arachnida, a class that includes spiders and other eight-leggers, they have four pairs of legs, a simple body type, and powerful jaws. Thousands of generations of evolution have pared their life to the straightforward: birth, mating, eating, and dying. Born legless and without sex organs, they soon grow these parts, mate, and climb to a high perch—say, a blade of grass. Here's where their tale gets striking. Of all the sights, sounds, and odors of the world, the adult tick is waiting for just one. It is not looking around: ticks are blind. No sound bothers the tick: sounds are irrelevant to its goal. It only awaits the approach of a single smell: a whiff of butyric acid, a fatty acid emitted by warm-blooded creatures (we sometimes smell it in sweat). It might wait here for a day, a month, or a dozen years. But as soon as it smells the odor it is fixed on, it drops from its perch. Then a second sensory ability kicks in. Its skin is photosensitive, and can detect warmth. The tick directs itself toward warmth. If it's lucky, the warm, sweaty smell is an animal, and the tick grasps on and drinks a meal of blood. After feeding once, it drops, lays eggs, and dies.

The point of this tale of the tick is that the tick's self-world is different than ours in unimagined ways: what it senses or wants; what its goals are. To the tick, the complexity of persons is reduced to two stimuli: smell and warmth—and it is very intent on those two things. If we want to understand the life of any animal, we need to know what things are meaningful to it. The first way to discover this is to determine what the animal can perceive: what it can see, hear, smell, or otherwise sense. Only objects that are perceived can have meaning to the animal; the rest are not even noticed, or all look the same. The wind that whisks through the grasses? Irrelevant to the tick. The sounds of a childhood birthday party? Doesn't appear on its radar. The delicious cake crumbs on the ground? Leave the tick cold.

Second, how does the animal act on the world? The tick mates, waits, drops, and feeds. So the objects of the universe, for the tick, are divided into ticks and non-ticks; things one can or cannot wait upon; surfaces one might or might not drop onto; and substances one may or may not want to feed on.

Thus, these two components—perception and action— largely define and circumscribe the world for every living thing. All animals have their own *umwelten*—their own subjective realities, what von Uexküll thought of as "soap bubbles" with them forever caught in the middle. We humans are enclosed in our own soap bubbles, too. In each of our self-worlds, for instance, we are very attentive to where other people are and what they are doing or saying. (By contrast, imagine the tick's indifference to even our most moving monologues.) We see in the visual range of light, we hear audible noises, and we smell strong odors placed in front of our noses. On top of that each individual creates his own personal umwelt, full of objects with special meaning to him. You can most clearly see this last fact by letting yourself be led through an unknown city by a native. He will steer you along a path obvious to him, but invisible to you. But the two of you share some things: neither of you is likely to stop and listen to the ultrasonic cry of a nearby bat; neither of you smells what the man passing you had for dinner last night (unless it involved a lot of garlic). We, the ticks, and every other animal dovetail into our environment: we are bombarded with stimuli, but only a very few are meaningful to us.

The same object, then, will be seen (or, better, *sensed*—some animals do not see well or at all) by different animals differently. A rose is a rose is a rose. Or is it? To a human a rose is a certain kind of flower, a gift between lovers, and a thing of beauty. To the beetle, a rose is perhaps an entire territory, with places to hide (on the underside of a leaf, invisible to aerial predators), hunt (in the head of the flower where ant nymphs grow), and lay eggs (in the joint of the leaf and stem). To the elephant, it is a thorn barely detectable underfoot.

And to the dog, what is a rose? As we'll see, this depends upon the construction of the dog, both in body and brain. As it turns out, to the dog, a rose is neither a thing of beauty nor a world unto itself. A rose is undistinguished from the rest of the plant matter surrounding it—unless it has been urinated upon by another dog, stepped on by another animal, or handled by the dog's owner. Then it gains vivid interest, and becomes far more significant to the dog than even the well-presented rose is to us.

PUTTING OUR UMWELT CAPS ON

Discerning the salient elements in an animal's world—his umwelt—is, in a sense, becoming an expert on the animal: whether a tick, a dog, or a human being. And it will be our tool for resolving the tension between what we think we know about dogs, and what they are actually doing. Yet without anthropomorphisms we would seem to have little vocabulary with which to describe their perceived experience.

Understanding a dog's perspective—through understanding his abilities, experience, and communication—provides that vocabulary. But we can't translate it simply through an introspection that brings our own umwelt along. Most of us are not excellent smellers; to imagine being a smeller, we have to do more than just think on it. That kind of introspective exercise only works when paired with an understanding of how profound the difference in umwelt is between us and another animal.

We can glimpse this by "acting into" the umwelt of another animal, trying to embody the animal—mindful of the constraints our sensory system places on our ability to truly do so. Spending an afternoon at the height of a dog is surprising. Smelling (even with our impoverished schnozes) every object we come across in a day closely and deeply yields a new dimension on otherwise familiar things. As you read this, try attending to all the sounds in the room you are in now that you have become accustomed to and usually tune out. With attention I suddenly hear the fan behind me, a beeping truck heading in reverse, the murmurations of a crowd of voices entering the building downstairs; someone adjusts their body in a wooden chair, my heart beats, I swallow, a page is turned. Were my hearing keener, I might notice the scratch of pen on paper across the room; the sound of a plant stretching in growth; the ultrasonic cries of the population of insects always underfoot. Might these noises be in the foreground in another animal's sensory universe?

THE MEANING OF THINGS

Even the objects in a room are not, in some sense, the *same objects* to another animal. A dog looking around a room does not think he is surrounded by human things; he sees *dog things*. What we think an object is for, or what it makes us think of, may or may not match the dog's idea of the object's function or meaning. Objects are defined by how you can act upon them: what von Uexküll calls their *functional tones*—as though an object's use rings bell-like when you set eyes on it. A dog may be indifferent to chairs, but if trained to jump on one, he learns that the chair has a *sitting tone:* it can be sat upon. Later, the dog might himself decide that other objects have a sitting tone: a sofa, a pile of pillows, the lap of a person on the floor. But other things that we identify as chairlike are not so seen by dogs: stools, tables, arms of couches. Stools and tables are in some other category of objects: obstacles, perhaps, in their path toward the *eating tone* of the kitchen.

Here we begin to see how the dog and the human overlap in our worldviews, and how we differ. A good many objects in the world have an eating tone to the dog—probably many more than we see as such. Feces just aren't menu items for us; dogs disagree. Dogs may have tones that we don't have at all—*rolling tones,* say: things that one might merrily roll in. Unless we are particularly playful or young, our list of rolling-tone objects is small to nil. And plenty of ordinary objects that have very specific meanings to us—forks, knives, hammers, pushpins, fans, clocks, on and on—have little or no meaning to dogs. To a dog, a hammer doesn't exist. A dog doesn't act with or on a hammer, so it has no significance to a dog. At least, not unless it overlaps with some other, meaningful object: it is wielded by a loved person; it is urinated on by the cute dog down the street; its dense wooden handle can be chewed like a stick.

A clash of umwelts occurs when dog meets human, and it tends to result in people misunderstanding what their dog is doing. They aren't seeing the world from the dog's perspective: the way he sees it. For instance, dog owners commonly insist, in grave tones, that a dog is never to lie on the bed. To drum in the seriousness of this dictum, this owner may go out and purchase what a pillow manufacturer has decided to label a "dog bed," and place it on the floor. The dog will be encouraged to come and lie on this special bed, the non-forbidden bed. The dog typically will do so, reluctantly. And thus one might feel satisfied: another dog-human interaction successful!

But is it so? Many days I returned home to find a warm, rumpled pile of sheets on my bed where either the wagging dog who greeted me at the door, or some unseen sleepy intruder, recently lay. We have no trouble seeing the meaning of the beds to the human: the very names of the objects make the situation clear. The big bed is for people; the dog bed is for dogs. Human beds represent relaxation, may be expensively outfitted with specially chosen sheets, and display all manner of fluffed pillows; the dog bed is a place we would never think to sit, is (relatively) inexpensive, and is more likely to be adorned with chew toys than with pillows. What about to the dog? Initially, there's not much difference between the beds—except, perhaps, that our bed is infinitely more desirable. Our beds smell like us, while the dog bed smells like whatever material the dog bed manufacturer had lying around (or, worse, cedar chips—overwhelming perfume to a dog but pleasant to us). And our beds are where *we* are: where we spend idle time, maybe shedding crumbs and clothes. The dog's preference? Indisputably our bed. The dog does not know all the things about the bed that make it such a glaringly different object to us. He may, indeed, come to learn that there is something different about the bed—by getting repeatedly scolded for lying on it. Even then, what the dog knows is less "human bed" versus "dog bed" but "thing one gets yelled at for being on" versus "thing one does not get yelled at for being on."

In the dog umwelt beds have no special functional tone. Dogs sleep and rest where they can, not on objects designated by people for those purposes. There may be a functional tone for places to sleep: dogs prefer places that allow them to lie down fully, where the temperature is desirable, where there are other members of their troop or family around, and where they are safe. Any flattish surface in your home satisfies these conditions. Make one fit these criteria, and your dog will probably find it just as desirable as your big, comfy human bed.

ASKING DOGS

To bolster our claims about the experience or mind of a dog, we will learn how to ask the dog if we're right. The trouble, of course, with asking a dog if he is happy or depressed is not that the question makes no sense. It's that we are very poor at understanding his response. We're made terribly lazy by language. I might guess at the reasons behind my friend's recalcitrant, standoffish behavior for weeks, forming elaborate, psychologically complex descriptions of what her actions indicate about what she thinks I meant on some fraught occasion. But my best strategy by leaps is to simply ask her. She'll tell me. Dogs, on the other hand, never answer in the way we'd hope: by replying in sentences, well punctuated and with italicized emphases. Still, if we look, they have plainly answered.

For instance, is a dog who watches you with a sigh as you prepare to leave for work depressed? Are dogs left at home all day pessimistic? Bored? Or just exhaling idly, preparing for a nap?

Looking at behavior to learn about an animal's mental experience is precisely the idea behind some cleverly designed recent experiments. The researchers used not dogs, but that shopworn research subject, the laboratory rat. The behavior of rats in cages may be the single largest contributor to the corpus of psychological knowledge. In most cases, the rat itself is not of interest: the research isn't about rats per se. Surprisingly, it's about humans. The notion is that rats learn and remember by using some of the same mechanisms that humans use—but rats are easier to keep in tiny boxes and subject to restricted stimuli in the hopes of getting a response. And the millions of responses by millions of laboratory rats, *Rattus norvegicus,* have greatly informed our understanding of human psychology.

But the rats themselves are intrinsically interesting as well. People who work with rats in laboratories sometimes describe their animals' "depression" or their exuberant natures. Some rats seem lazy, some are cheery; some pessimistic, some optimistic. The researchers took two of these characterizations—pessimism and optimism—and gave them operational definitions: definitions in terms of behavior that allow us to determine whether real differences in the rats can be seen. Instead of simply extrapolating from how humans look when pessimistic, we can ask how a pessimistic rat might be distinguished by its behavior from an optimistic one.

Thus, the rats' behavior was examined not as a mirror to our own but as indicating something about . . . rats: about rat preference and rat emotions. Their subjects were placed in tightly restricted environments: some were "unpredictable" environments, where the bedding, cage mates, and the light and dark schedule were always changing; others were stable, predictable environments. The experimental design took advantage of the fact that, hanging out in their cages with little to do, rats quickly learn to associate new events with simultaneously occurring phenomena. In this case, a particular pitch was played over speakers into the cages of the rats. It was a prompt to press a lever: the lever triggered the arrival of a pellet of food. When a different pitch was played and the rats pressed the lever, they were greeted with an unpleasant sound and no food. These rats, reliably like lab rats before them, quickly learned the association. They raced over to the food-dispensing lever only when the good-harbinger sound appeared, like young children rallied by the jingle of an ice-cream truck. All of the rats learned this easily. But when the rats were played a new sound, one between the two learned pitches, what the researchers found was that the rats' environment mattered. Those who had been housed in a predictable environment interpreted the new sound to mean *food;* those in unstable environments did not.

These rats had learned optimism or pessimism about the world. To watch the rats in the predictable environments jump with alacrity at every new sound is to see optimism in action. Small changes in the environment were enough to prompt a large change in outlook. Rat lab workers' intuitions about the mood of their charges may be spot-on.

We can subject our intuitions about dogs to the same kind of analysis. For any anthropomorphism we use to describe our dogs, we can ask two questions: One, is there a natural behavior this action might have evolved from? And two, what would that anthropomorphic claim amount to if we deconstructed it?

DOG KISSES

> Licks are Pump's way of making contact, her hand outstretched
> for me. She greets me home with licks at my face as I bend to pet
> her; I get waking licks on my hand as I nap in a chair; she licks
> my legs thoroughly clean of salt after a run; sitting beside me, she
> pins my hand with her front leg and pushes open my fist to lick
> the soft warm flesh of my palm. I adore her licks.

I frequently hear dog owners verify their dogs' love of them through the kisses delivered upon them when they return home. These "kisses" are licks: slobbery licks to the face; focused, exhaustive licking of the hand; solemn tongue-polishing of a limb. I confess that I treat Pump's licks as a sign of affection. "Affection" and "love" are not just the recent constructs of a society that treats pets as little people, to be shod in shoes in bad weather, dressed up for Halloween, and indulged with spa days. Before there was any such thing as a doggy day care, Charles Darwin (who I feel confident never dressed up his pup as a witch or goblin) wrote of receiving lick-kisses from his dogs. He was certain of their meaning: dogs have, he wrote, a "striking way of exhibiting their affection, namely, by licking the hands or faces of their masters." Was Darwin right? The kisses feel affectionate *to me,* but are they gestures of affection *to the dog*?

First, the bad news: researchers of wild canids—wolves, coyotes, foxes, and other wild dogs—report that puppies lick the face and muzzle of their mother when she returns from a hunt to her den—in order to get her to regurgitate for them. Licking around the mouth seems to be the cue that stimulates her to vomit up some nicely partially digested meat. How disappointed

Pump must be that not a single time have I regurgitated half-eaten rabbit flesh for her.

Furthermore, our mouths taste great to dogs. Like wolves and humans, dogs have taste receptors for salty, sweet, bitter, sour, and even umami, the earthy, mushroomy-seaweedy flavor captured in the flavor-heightening monosodium glutamate. Their perception of sweetness is processed slightly differently than ours, in that salt enhances the experience of sweet tastes. The sweet receptors are particularly abundant in dogs, although some sweeteners—sucrose and fructose—activate the receptors more than others, such as glucose. This could be adaptive in an omnivore like the dog, for whom it pays to distinguish between ripe and non-ripe plants and fruit. Interestingly, even pure salt doesn't kick-start the so-called salt receptors on the tongue and the roof of the mouth in dogs the way it does in humans. (There's some disagreement whether dogs have salt-specific receptors at all.) But it didn't take long reflecting on her behavior for me to realize that Pump's licks to my face often correlated with my face having just overseen the ingestion of a good amount of food.

Now the good news: as a result of this functional use of mouth licking—"kisses" to you and me—the behavior has become a ritualized greeting. In other words, it no longer serves only the function of asking for food; now it is used to say hello. Dogs and wolves muzzle-lick simply to welcome another dog back home, and to get an olfactory report of where the home-comer has been or what he has done. Mothers not only clean their pups by licking, they often give a few darting licks when reuniting after even a brief time apart. A younger or timid dog may lick the muzzle, or muzzle vicinity, of a bigger, threatening dog to appease him. Familiar dogs may exchange licks when meeting at their ends of their respective leashes on the street. It may serve as

a way to confirm, through smell, that this dog storming toward
them is who they think he is. Since these "greeting licks" are often
accompanied by wagging tails, mouths opened playfully, and
general excitement, it is not a stretch to say that the licks are a way
to express happiness that you have returned.

DOGOLOGIST

I still talk about Pump's looking "knowingly," or feeling *content*
or *capricious*. These are words that capture something to me. But
I have no illusion that they map to her experience. And I still
adore her licks; but I also adore knowing what they mean to her
rather than just what they mean to me.

By imagining the umwelt of dogs, we'll be able to decon-
struct other anthropomorphisms—of our dog's guilt at chewing
a shoe; of a pup's revenge wrought on your new Hermès scarf—
and reconstruct them with the dog's understanding in mind.
Trying to understand a dog's perspective is like being an anthropologist in a foreign land—one peopled entirely by dogs. A
perfect translation of every wag and woof may elude us, but simply looking closely will reveal a surprising amount. So let's look
closely at what the natives do.

In the following chapters we will consider the many dimensions contributing to a dog's umwelt. The first dimension is
historical: how dogs came from wolves, and how they are and
are not wolflike. The choices we've made in breeding dogs led to
some intentional designs and some unintended consequences.
The next dimension comes from anatomy: the dog's sensory
capacity. We need to appreciate what the dog smells, sees, and
hears . . . and if there are other means by which to sense the
world. We must imagine the view from two feet off the ground,

and from behind such a snout. Finally, the body of the dog leads us to the brain of the dog. We'll look at the dog's cognitive abilities, the knowledge of which can help us to translate their behavior. Together, these dimensions combine to provide answers to the questions of what dogs think, know, and understand. Ultimately they will serve as scientific building blocks for an informed imaginative leap inside of a dog: halfway to being honorary dogs ourselves.

Belonging to the House

She's waiting at the threshold of the kitchen, just beyond underfoot. Somehow Pump knows precisely where "out of the
kitchen" is. Here she sprawls, and when I bring food to the
table, she ducks in to retrieve the kitchen fallings. At the table,
she gets a little of everything—and she'll at least entertain even
the most unlikely offering, if only to loll it around in her mouth
before unceremoniously depositing it on the ground. She does
not like raisins. Nor tomatoes. She'll suffer a grape, if she man-
ages to split it into juicy halves with her side teeth—then delib-
erating, as though managing a very big or tough object, and
masticating it. All carrot ends are for her. She takes the stems of
broccoli and asparagus and holds them gently, gazing at me for
a moment as if determining if anything else is coming before
walking to the rug to settle down for a gnaw.

Dog training books often insist that "a dog is an animal": this is
true but is not the whole truth. The dog is an animal *domesticated,* a word that grew from a root form meaning "belonging to
the house." Dogs are animals who belong around houses.
Domestication is a variation of the process of evolution, where

the selector has been not just natural forces but human ones, eventually intent on bringing dogs inside their homes.

To understand what the dog is about we have to understand from where he came. As a member of the *Canidae* family—all of whose members are called *canids*—the domestic dog is distantly related to coyotes and jackals, dingoes and dholes, foxes and wild dogs.* But he arose from just one ancient *Canidae* line, animals most likely resembling the contemporary gray wolf. When I see Pumpernickel delicately spit out a raisin, though, I am not reminded of the stark images of wolves in Wyoming downing a moose and yanking it apart.† The existence of an animal who will patiently wait at the kitchen door, and then ponderously consider a carrot stick, seems at first glance irreconcilable with that of an animal whose primary allegiance is to himself, whose affiliations are fraught with tension and maintained by force.

Carrot-considerers arose out of moose killers through the second source: us. Where nature blindly, uncaringly "selects" traits that lead to the survival of their bearers, ancestral humans have also selected traits—physical features and behaviors—that have led not just to the survival, but to the omnipresence of the modern dog, *Canis familiaris,* among us. The animal's appearance, behavior, preferences; his interest in us and attention to our attention: these are largely the result of domestication. Present-day dog is a well-designed creature. Only much of this design was utterly unintentional.

---

*Not on this list are *hyenas*. Dog-sized and -shaped, with erect German shepherd–like ears, and prone to howl and vocalize like many garrulous canids, hyenas are in some ways doglike, but are not in fact canids. They are carnivores more closely related to mongooses and cats than to dogs.

†Raisins—and grapes— are now suspected of being toxic to some dogs, even in small amounts (though the mechanism of toxicity is unknown)—leading me to wonder whether Pump was instinctively averse to raisins.

HOW TO MAKE A DOG:
STEP-BY-STEP INSTRUCTIONS

So you want to make a dog? There are just a few ingredients. You'll need wolves, humans, a little interaction, mutual tolerance. Mix thoroughly and wait, oh, a few thousand years.

Or, if you're the Russian geneticist Dmitry Belyaev, you simply find a group of captive foxes and start selectively breeding them. In 1959, Belyaev began a project that has greatly informed our best guesses as to what we believe the earliest steps of domestication were. Instead of observing dogs and extrapolating backward, he examined another social canid species and propagated them forward. The silver fox in Siberia in the mid-twentieth century was a small, wild animal that had become popular with the fur trade. Kept in pens, bred for their choice fur coats, particularly long and soft, the fox was not tamed but was captive. What Belyaev made of them, with a much reduced recipe, were not "dogs," but were surprisingly close to dogs.

Though *Vulpes vulpes,* the silver fox, is distantly related to wolves and dogs, it had never before been domesticated. Despite their evolutionary relatedness, no canids are fully domesticated other than the dog: domestication doesn't happen spontaneously. What Belyaev showed was that it can happen quickly. Beginning with 130 foxes, he selectively chose and bred those that were the most "tame," as he described it. What he really chose were those foxes that were the least fearful of or aggressive toward people. The foxes were caged, so aggression was minimal. Belyaev approached each cage and invited the fox to eat some food out of his hand.

Some bit at him; some hid. Some took the food, reluctantly. Others took the food and also let themselves be touched and patted without fleeing or snarling. Still others accepted the food and even wagged and whimpered at the experimenter, inviting rather than discouraging interaction. These were the foxes Belyaev selected. By some normal variation in their genetic code, these animals were naturally calmer around people, even interested in people. None of them had been trained; all had the same, minimal exposure to human caretakers, who fed them and cleaned their bedding for their short lives.

These "tame" foxes were allowed to mate, and their young were tested the same way. The tamest of those were mated, when they were old enough; and their young; and their young. Belyaev continued the work until his death, and the program has continued since. After forty years, three-quarters of the population of foxes were of a class the researchers called "domesticated elite": not just accepting contact with people, but drawn to it, "whimpering to attract attention and sniffing and licking" . . . as dogs do. He had created a domesticated fox.

Later genomic mapping has revealed that forty genes now differ between Belyaev's tame foxes and the wild silver fox. Incredibly, by selecting for one behavioral trait, the genome of the animal was changed in a half century. And with that genetic change came a number of surprisingly familiar physical changes: some of the later-generation foxes have multicolored, piebald coats, recognizable in dog mutts everywhere. They have floppy ears and tails that curl up and over their backs. Their heads are wider and their snouts are shorter. They are improbably cute.

All these physical characteristics came along for the ride, once a particular behavior was chosen and picked out. The behavior is not what affects the body; instead, both are the common result of a gene or set of genes. Single behaviors aren't dictated by genes, but they are made more or less likely by them. If someone's genetic makeup leads to having very high levels of a stress hormone, for instance, it doesn't mean that they will be stressed all the time. But it may mean that they have a lowered threshold for having the classic stress response—a raised heart rate and breathing rate, increased sweating, and so on—in some contexts where someone else doesn't have a stress response. Let's say this low-threshold character screams at her dog for barreling into her at the dog park. Her screaming at the poor pup certainly is not genetically obliged—genes don't know from dog parks, or even pups—but her neurochemistry, created from her genes, facilitated it happening when a situation presented itself.

So, too, with the doglike foxes. Given what genes do,* even a small change in a gene—turning on slightly later than it otherwise might, say—could change the likelihood of both certain behaviors and certain forms of physical appearance. Belyaev's foxes show that a few simple developmental differences can have a wide-ranging effect: for instance, his foxes open their eyes earlier and show their first fear responses later, more like dogs than wild foxes. This gives them a longer early window for bonding with a caretaker—such as a human experimenter in Siberia. They play with each other even when they reach adulthood, perhaps allowing for longer and more complex socialization. It is worth noting that foxes diverged from wolves some ten to twelve million years ago; yet in forty years' of selection they look domesticated. The same perhaps could happen with other carnivores we take under our wing and inside our houses. The genetic changes nudge them into being doggy.

---

*What (some) genes do is regulate the formation of proteins that assign cells their roles. When, where, and in what environment a cell develops all contribute to the result. Thus the path from a gene to the emergence of a physical trait or a behavior is more circuitous than one might initially think, with room for modifications along the way.

HOW WOLVES BECAME DOGS

Though we tend not to think much about it, the history of dogs, well before you got your dog, bears more on what your dog is like than the particulars of his parentage. Their history begins with wolves.

Wolves are dogs before the accoutrements. The coat of domestication makes dogs quite different creatures, however.* While a pet dog gone missing may not survive even a handful of days on his own, the anatomy, instinctual drive, and sociality of the wolf combine to make it very adaptable. These canids can be found in diverse environments: in deserts, forests, and on ice. For the most part, wolves live in packs, with one mating pair and from four up to forty younger, usually related wolves. The pack works cooperatively, sharing tasks. Older wolves may help raise the youngest pups, and the whole group works together when hunting large prey. They are very territorial and spend a good amount of time demarcating and defending their borders.

Inside some of these borders, tens of thousands of years ago, human beings began to appear. *Homo sapiens,* having outgrown his *habilis* and *erectus* forms, was becoming less nomadic and beginning to create settlements. Even before agriculture began, interactions between humans and wolves began. Just how those interactions played out is the source of speculation. One idea is

---

*There is some debate over whether dogs should be considered a separate species from, or a subspecies of wolves. There is even debate over whether the original Linnaean classification scheme that demarcates *species* as a fundamental unit is still helpful or valid. Most researchers agree that describing wolves and dogs as separate species is the best current description. Although the two animals can interbreed, their typical mating habits, their social ecology, and the environments they live in are very different.

that the humans' relatively fixed communities produced a large amount of waste, including food waste. Wolves, who will scavenge as well as hunt, would have quickly discovered this food source. The most brazen among them may have overcome any fear of these new, naked human animals and begun feasting on the scraps pile. In this way, an accidental natural selection of wolves who are less fearful of humans would have begun.

Over time, humans would tolerate the wolves, maybe taking a few pups in as pets, or, in leaner times, as meat. Generation by generation, the calmer wolves would have more success living on the edge of human society. Eventually, people would begin intentionally breeding those animals they particularly liked. This is the first step of domestication, a remaking of animals to our liking. With all species, this process typically occurs through a gradual association with humans, whereby successive generations become more and more tame and finally become distinct in behavior and body from their wild ancestors. Domestication is thus preceded by a kind of inadvertent selection of animals who are nearby, useful, or pleasing, allowing them to loiter on the edges of human society. The next step in the process involves more intention. Those animals who are less useful or liked are abandoned, destroyed, or deterred from hanging about with us. In this way, we select those animals who more easily submit to our breeding of them. Finally, and most familiar, domestication involves breeding animals for specific characteristics.

Archeological evidence dates the first domesticated wolf-cum-dog at ten thousand to fourteen thousand years ago. Dog remains have been found in trash heaps (suggesting their use as food or property) and in grave sites, their skeletons curled up aside human skeletons. Most researchers think dogs began to associate with us even earlier, maybe many tens of thousands of years ago. There is genetic evidence, in the form of mitochondrial DNA samples,* of a subtle split as long as 145,000 years ago between pure wolves and those that were to become dogs. We could call the latter wolves protodomesticators, since they had themselves changed behaviorally in ways that would later encourage humans' interest (or merely tolerance) of them. By the time humans came along, they might have been ripe for domesticating. The wolves taken up by humans were probably less hunters than scavengers, less dominant and smaller than alpha wolves, and tamer. In sum, less wolfy. Thus, early in the development of ancient civilizations, thousands of years before domesticating any other animal, humans took this one animal with them inside the walls of their fledgling villages.

These vanguard dogs would not be mistaken as members of one of the hundreds of currently recognized dog breeds. The short stature of the dachshund, the flattened nose of the pug— these are the results of selective breeding by humans much later. Most dog breeds we recognize today have only been developed in the last few hundred years. But these early dogs would have inherited the social skills and curiosity of their wolf ancestors, and would then have applied them toward cooperating with and appeasing humans as much as toward each other. They lost some of their tendency toward pack behavior: scavengers don't need the proclivity to hunt together. Nor is any hierarchy relevant when you might live and eat on your own. They were sociable but not in a social hierarchy.

The change from wolf to dog was striking in its speed. Humans took nearly two million years to morph from *Homo*

---

*Mitochondrial DNA are chains of DNA within the energy-producing mitochondria of cells, but outside the cell nucleus. They are inherited, without any change, from the mother by her offspring. The mtDNA of individuals has been used to trace human ancestry, and to estimate the evolutionary relationships among animal species.

habilis to *Homo sapiens,* but the wolf leapfrogged into dogness in a fraction of the time. Domestication mirrors what nature, through natural selection, does over hundreds of generations: a kind of artificial selection that hurries up the clock. Dogs were the first domesticated animals, and in some ways the most surprising. Most domestic animals are not predators. A predator seems like an unwise choice to take into one's home: not only would it be difficult to find provisions for a meat eater, one risks being seen as meat oneself. And though this might make them (and has made them) good hunting pals, their main role in the last hundred years has been to be a friend and non-judgmental confidant, not a worker.

But wolves do have features that made them terrific candidates for artificial selection. The process favors a social animal who is behaviorally flexible, able to adjust its behavior in different settings. Wolves are born into a pack, but only stay until they are a few years old: then they leave and find a mate, create a new pack, or join an already existing pack. This kind of flexibility to changing status and roles is well suited to dealing with the new social unit that includes humans. Within a pack or moving between packs, wolves would need to be attentive to the behavior of packmates—just as dogs will need to be attentive to their keepers and sensitive to their behavior. Those early wolf-dogs meeting early human settlers would not have benefited the humans much, so they must have been valued for some other reason—say, for their companionship. The openness of these canids allowed them to adjust to a new pack: one that would include animals of an entirely different species.

UNWOLFY

And so some wolflike ancestor of both wolves and dogs took the plunge, loitered among human loiterers, and was eventually adopted and then molded by humans instead of solely by the caprice of nature. This makes present-day wolves an interesting comparison species to dogs: they likely share many traits. The present-day wolf is not the ancestor of the dog; though wolves and dogs share a common ancestor. Even the modern wolf is likely quite different than the ancestral wolves. What is different between dogs and wolves is probably due to what made some protodogs likely to be taken in, plus whatever humans have done in breeding them since.

And there are many differences. Some are developmental: for instance, dogs' eyes don't open for two or more weeks, whereas wolf pups open their eyes at ten days old. This slight difference can have a cascading effect. Generally, dogs are slower to develop physically and behaviorally. The big developmental milestones—walking, carrying objects in the mouth, when they first engage in biting games—come generally later for dogs than for wolves.* This small difference blossoms into a large difference: it means that the window for socialization is different in dogs and wolves. Dogs have more leisure to learn about others and to become accustomed to objects in their environment. If dogs are exposed to non-dogs—humans or monkeys or rabbits or cats—in the first few months of development, they form an attachment to and prefer-

---

*There is also a large breed difference. For instance, poodles don't show avoidance behaviors or begin to play-fight until weeks after huskies do—when weeks represent a goodly chunk of the puppy's life. In fact, huskies develop more quickly than wolves in some ways. No one has studied how this affects their rapport with humans.

ence for these species over others, often trumping any predatory or fearful drive we might expect them to feel. This so-called *sensitive* or *critical* period of social learning is the time during which dogs will learn who is a dog, an ally, or a stranger. They are most susceptible to learning who their peers are, how to behave, and associations between events. Wolves have a smaller window during which to determine who is familiar and who is foe.

There are differences in social organization: dogs do not form true packs; rather, they scavenge or hunt small prey individually or in parallel.* Though they don't hunt cooperatively, they are cooperative: bird dogs and assistance dogs, for instance, learn to act in synchrony with their owners. For dogs, socialization among humans is natural; not so for wolves, who learn to avoid humans naturally. The dog is a member of a human social group; its natural environment, among people and other dogs. Dogs show what is called with human infants "attachment": preference for the primary caregiver over others. They have anxiety at separation from the caregiver, and greet her specially on her return. Though wolves greet other members of the pack when they reunite after being apart, they don't seem to show attachment to particular figures. For an animal who is going to be around humans, specific attachments make sense; for an animal who lives in a pack, it is less applicable.

Physically, dogs and wolves differ. While still quadrupedal omnivores, the range of body types and sizes among dogs is extraordinary. No other canid, or other species, shows the same diversity of body types within a species, from the four-pound

---

*As the domestication process probably began with early canids scavenging around human groups—eating our table scraps—it is a particularly silly stance to feed dogs only raw meat, on the theory that they are wolves at heart. Dogs are omnivores who for millennia have eaten what we eat. With very few exceptions, what is good on my plate is good for my dog's bowl.

papillon to the two-hundred-pound Newfoundland; from slender dogs with long snouts and whiplike tails to pudgy dogs with foreshortened noses and stubs of tails. Limbs, ears, eyes, nose, tail, fur, haunches, and belly are all dimensions along which dogs can be reconfigured and still be dogs. Wolves' sizes, by contrast, are, like most wild animals, fairly reliably uniform in a particular environment. But even the "average" dog—something resembling a prototypical mutt—is distinguishable from the wolf. The dog's skin is thicker than wolves'; while both have the same number and kind of teeth, the dog's are smaller. And the whole head is smaller on a dog than on a wolf: about 20 percent smaller. In other words, between a dog and a wolf of similar body size, the dog has the much smaller skull—and, correspondingly, a smaller brain.

This latter fact has continued to be promulgated, perhaps an indication of the ongoing appeal of the claim (now debunked) that brain size determines intellect. While erroneous, the smoothness of the shift from talking about brain *size* to brain *quality* trumped evidence to the contrary. Comparative studies with wolves and dogs on problem-solving tasks initially seemed to confirm dogs' cognitive inferiority. Hand-raised wolves tested on their ability to learn a task—to pull three ropes from an array of ropes in a particular order—well outperformed the dogs tested. The wolves more quickly learned to pull any rope to begin and then proceeded to be more successful at learning the order in which the ropes were to be pulled. (They also tore more ropes to pieces than the dogs did, although the researchers are silent about what this indicates about their cognition.) Wolves are also great at escaping from enclosed cages; dogs are not. Most canid researchers agree that wolves pay more attention than dogs do to physical objects and handle these objects more capably.

From results like these comes the notion that there is a cognitive difference between wolves and dogs: usually, that wolves are insightful problem solvers, and dogs simpletons. In actual fact, historically theories have oscillated between claiming dogs to be more intelligent, or wolves to be the smarter of the two. Science is often contingent on the culture in which it is practiced, and these theories reflect the then-prevailing ideas about animal minds. The accumulated data of dog and wolf behavior, though, leads to a more nuanced position. Wolves seem to be better at solving certain kinds of *physical* puzzles. Some of this skill is explainable by looking at their natural behavior. Why did wolves readily learn the rope-pulling task? Well, they do a lot of grabbing and pulling on things (like prey) in their natural environment. Some of the difference can be traced to dogs' more limited requirements for living. Having been folded into the world of humans, dogs no longer need some of the skills that they would to survive on their own. As we'll see, what dogs lack in physical skills, they make up for in people skills.

AND THEN OUR EYES MET . . .

There is a final, seemingly minor difference between the two species. This one small behavioral variation between wolves and dogs has remarkable consequences. The difference is this: dogs look at our eyes.

Dogs make eye contact and look to us for information— about the location of food, about our emotions, about what is happening. Wolves avoid eye contact. In both species, eye contact can be a threat: to stare is to assert authority. So too is it with humans. In one of my undergraduate psychology classes, I have my students do a simple field experiment wherein they try to make and hold eye contact with everyone they pass on campus.

Both they and those on the receiving end of their stares behave remarkably consistently: everyone can't wait to break eye contact. It's stressful for the students, a great number of whom suddenly claim to be shy: they report that their hearts begin to race and they start sweating when simply holding someone's gaze for a few seconds. They concoct elaborate stories on the spot to explain why someone looked away, or held their gaze for a half second longer. For the most part, their staring is met with deflected gazes from those they eyeball. In a related experiment, they test gaze in a second way, verifying our species' tendency to follow the gaze of others to its focal point. A student approaches any publicly visible and shared object—a building, tree, spot on the sidewalk—and looks fixedly at one point on it. Her partner, another student, stands nearby and surreptitiously records the reactions of passersby. If it's not rush hour and raining, they report finding that at least some people stop in their tracks to follow their gaze and stare curiously at that fascinating sidewalk spot: surely there must be *something*.

If this behavior is unsurprising, it is because it is so human: we look. Dogs look, too. Though they have inherited some aversion to staring too long at eyes, dogs seem to be predisposed to inspect our faces for information, for reassurance, for guidance. Not only is this pleasing to us—there is a certain satisfaction in gazing deep into a dog's eyes gazing back at you—it is also perfectly suited to getting along with humans. As we will see later in this book, it also serves as a foundation for their skill at social cognition. We not only avoid eye contact with strangers, we rely on eye contact with intimates. There is information in a furtive glance; a gaze mutually held feels profound. Eye contact between people is essential to normal communication.

Hence a dog's ability to find and gaze at our eyes may have been one of the first steps in the domestication of dogs: we chose those that looked at us. What we then did with dogs is peculiar. We began designing them.

FANCY DOGS

The label on her cage said "Lab mix." Every dog in the shelter was a Lab mix. But surely Pump was born of a spaniel: her black, silky hair fell against her slender frame; her velvet ears framed her face. In sleep she was a perfect bear cub. Soon her tail hairs grew longer and feathery: so she's a golden retriever. Then the gentle curls on her underbelly tightened; her jowls filled a bit: okay, she's a water dog. As she ages her belly grows until she has a solid, barrel-like shape—she's a Lab after all; her tail becomes a flag needing trimming—a Lab/golden mix; she could be still one moment and sprinting the next—a poodle. She is curly and round-bellied: clearly the product of a sheepdog who'd snuck into the bushes with a pretty sheep. She's her own dog.

The original dogs were mongrels, in the sense that they didn't come from a controlled lineage. But many of the dogs we keep, mutts or not, emerged from hundreds of years of strictly controlled breeding. The consequence of this breeding is the creation of what are nearly subspecies, varying in shape, size, lifetime, temperament,* and skills. The outgoing Norwich ter-

---

Temperament is used to mean roughly the same thing as *personality,* without the overtone of anthropomorphism. It is perfectly acceptable to talk about a dog's personality, if we mean the dog's "usual pattern of behavior and individual traits": behavior and traits are not exclusive to humans. Some researchers use *temperament* to refer to the traits as they appear in a young animal—the genetic tendency of the dog; while reserving *personality* to refer to adult traits and behaviors, the result of that particular temperament combined with whatever they confronted in their environment.

rier, ten inches high and ten pounds strong, is but the weight of the calm, sweet, enormous Newfoundland's head. Ask some dogs to retrieve a ball and you get a puzzled look; but a border collie doesn't need to be asked twice.

The familiar differences between modern breeds aren't always the result of intentional selection. Some behaviors and physical features are selected for—retrieving prey, smallness, a tightly curled tail—and some just come along for the ride. The biological reality of breeding is that the genes for traits and behaviors come in clusters. Mate a few generations of dogs with particularly long ears and you might find that they all share other characteristics: a strong neck, downcast eyes, fine jowls. Coursing dogs, bred to gallop swiftly or long, are leggy—their leg length matches (in the husky) or surpasses (in the greyhound) the depth of their chests. By contrast, dogs who track on the ground (as the dachshund) wound up with legs much shorter than their chest is deep. Similarly, selecting for one particular behavior inadvertently selects for accompanying behaviors. Breed dogs who are very sensitive to motion—who probably have an overabundance of rod photoreceptors in their retinae—and you may also get a dog whose acute sensitivity to motion leads to their being temperamentally high-strung. Their appearance might change, too: they may have large, globular eyes for seeing at night. Sometimes what comes to be desirable in a breed is a trait that first appeared inadvertently.

There is evidence of distinct dog breeds as early as five thousand years ago. In drawings from ancient Egypt, at least two kinds of dogs are depicted: mastiff-looking dogs, big of head and body, and slim dogs with curled tails.* The mastiffs may have

---

*There is no evidence, however, that any currently existing breed can lay claim to being the descendents of the original breeds. Descriptions of both the Pharaoh and

been guard dogs; the slender dogs appear to have been hunting companions. And so the designing of dogs for particular purposes began—and continued along these lines for a long while. By the sixteenth century, there were added other hounds, bird dogs, terriers, and shepherds. By the nineteenth century, clubs and competitions sprouted, and the naming and monitoring of breeds exploded.

The various modern breeds have probably all emerged with this proliferation of breeding in the last four hundred years. The American Kennel Club now lists nearly one hundred fifty varieties, grouped according to the purported* occupation of the breed. Hunting companions are distributed into "sporting," "hound," "working," and "terrier" categories; there are, in addition, the working "herding" breeds, the plainly "non-sporting" breeds, and the rather self-explanatory "toys." Even among dogs bred to join the hunt, there are subdivisions, by the very kind of assistance they provide (pointers point out the prey; retrievers retrieve it; Afghan hounds exhaust it); by the specific prey they're after (terriers are ratters, and harriers go after hares); and by the medium preferred (beagles chase on land; spaniels will swim in water). Worldwide, there are hundreds more breeds still. Breeds vary not just by our uses of them but physically: by body size, head size, head shape, body shape, type of tail, coat kind, coat color. Go searching for a purebred dog and you'll confront a new-

---

Ibizan hounds cite them as the "oldest" dog breeds, which claims seemed supported by their physical resemblance to the dogs of Egyptian paintings. However, their genomes reveal them to have emerged much more recently.

*The named occupation is mostly theoretical because a minority of dogs bred for work actually do the work of their breed (predominantly hunting or herding). The rest wind up either as companions sitting on our laps, or trained, trimmed, and blown-dry to be shown at "dog fancy" shows—odd, as nibbling the crusts of our sandwiches after a nice shampoo is about as far as one can get from dredging fallen waterfowl from a swamp.

car-worthy list of specs, detailing everything from the ears to the temperament of your future pup. Want a long-limbed, short-haired, jowly dog? Consider the Great Dane. In more of a short-nosed, rolled-skin, curly-tailed kind of mood? Here's a nice pug for you. Choosing between breeds is like choosing between anthropomorphized option packages. You not only get a dog, you get one who is typically "dignified, lordly, scowling, sober and snobbish" (shar-pei); "merry and affectionate" (English cocker spaniel); "reserved and discerning with strangers" (chow chow); with a "rollicking personality" (Irish setter); full of "self-importance" (Pekingese); having "heedless, reckless pluck" (Irish terrier); "equable" (Bouvier des Flandres); or, most surprisingly, "a dog at heart" (briard).

The dog fanciers will be surprised to hear, perhaps, that the grouping of breeds based on genetic similarity does not result in the same groupings as the AKC. Cairn terriers are closer to hounds; shepherds and mastiffs share much of their genomes. The genome belies most people's assumptions about dogs' similarities to wolves, too: the long-haired, sickle-tailed huskies are closer to wolves than the long-bodied, skulking German Shepherd. Basenjis, who bear almost no physical resemblance to wolves, are closer still. This is yet another indication that, for most of their domestication, the dog's appearance was an accidental side effect of his breeding.

Dog breeds are relatively closed genetic populations, meaning that each breed's gene pool is not accepting new genomes from outside the pool. To be a member of a breed, a dog must have parents who are themselves members. Thus any physical changes in the offspring can only come from random genetic mutations, not from the mixing of different gene pools that usually occurs when animals (including humans) mate. Mutations, variation, and admixtures are generally good for populations, though, and help to prevent inherited disease: this is why purebred dogs, though they come from what is considered "good stock" in that the ancestry of the dogs is traceable through the breeding line, are more susceptible to many physical disorders than are mixed-breed dogs.

One boon of a closed gene pool is that the genome of a breed can be mapped, and in fact it recently has: a boxer's genome was the first, around nineteen thousand genes' worth. As a result, scientists are starting to make an accounting of where on the genome the genetic variations are that lead to characteristic traits and disorders, such as narcolepsy, the sudden and total fall into unconsciousness to which some dog breeds (particularly Dobermans) are susceptible.

Another advantage of a closed gene pool of a breed discussed by researchers is that it feels as though one is getting a relatively reliable animal when one selects from it. One can pick a "family-friendly" dog or one advertised as a skillful house guardian. But it is not so simple: dogs, like us, are more than their genome. No animal develops in a vacuum: genes interact with the environment to produce the dog you come to know. The exact formulation is difficult to specify: the genome shapes the dog's neural and physical development, which itself partially determines what will be noticed in the environment—and whatever is noticed itself further shapes continued neural and physical development. As a result, even with inherited genes, dogs aren't just carbon copies of their parents. On top of this, there is also great natural variability in the genome. Even a cloned dog, should you be tempted to replicate your beloved pet, will not be identical to the original: what a dog experiences and whom he meets will influence who he becomes in innumerable, untraceable ways.

So although we have tried to design dogs, the dogs we see today are partly creatures of serendipity. *What breed is she?* is a question I've been asked about Pump more than any other—and I in turn ask of other dogs. Her mongrelness encourages the great game of guessing her heritage: the resulting hunches are satisfying, even though none could ever be verified.*

THE ONE DIFFERENCE BETWEEN BREEDS

Though there is an extensive literature of dog breeds, there has never been a scientific comparison of breed behavior differences: a comparison that controls each animal's environment, giving them the same physical objects, the same exposure to dogs and humans, the same everything. It's hard to believe, given that such bold statements are made about what each breed is like. This is not to suggest that the differences are minimal or nonexistent. Dogs of various breeds will doubtless behave differently when, say, they are presented with a nearby, running rabbit. But it would be a mistake to guarantee that a dog, bred or not, will inevitably act a certain way on seeing that rabbit. This

---

*Genetic analysis tests have become available since the mapping of the genome: for a fee, companies will allegedly resolve your dog's genetic code, determined from a blood sample or swab of cheek cells, into its contributing breeds. At present the accuracy of the tests is indeterminate.

is the same mistake that is made when we wind up calling some breeds "aggressive" and legislating against them.*

Even without knowing the specific differences in the Labrador retriever's and the Australian shepherd's reaction to that rabbit, there is one thing that may account for the variability in behavior between breeds. They have different *threshold levels* to notice and react to stimuli. The same rabbit, for instance, causes different amounts of excitement in two different dogs; similarly, the same amount of hormone producing that excitement causes different rates of response, from raising a head in mild interest, to a full-on chase.

There is a genetic explanation behind this. Though we call a dog a retriever or a shepherd, it is not the behavior *retrieving* or *shepherding* that was selected for. Instead, it was the likelihood that the dog would respond just the right amount to various events and scenes. However, there is no one gene we can point to here. No gene develops right into *retrieving* behavior—or into any particular behavior at all. But a set of genes may affect the likelihood that an animal acts in a certain way. In humans, too, a genetic difference between individuals may appear as different propensities to certain behaviors. One might be more or less susceptible to becoming addicted to stimulant drugs, based partly on how much stimulation one's brain needs to produce a pleasurable feeling. Addictive behavior is thereby traceable to

*What is considered *aggressive* is culturally and generationally relative. German shepherds were on the top of the list after World War II; in the 1990s Rottweilers and Dobermans were scorned; the American Staffordshire terrier (also known as the pit bull) is the current bête noire. Their classification has more to do with recent events and public perception than with their intrinsic nature. Recent research found that of all breeds, *dachshunds* were the most aggressive to both their own owners and to strangers. Perhaps this is underreported because a snarling dachshund can be picked up and stashed away in a tote bag.

genes that design the brain—but there is no gene for *addiction*. The environment is clearly important here, too. Some genes regulate expression of other genes—which expression might depend on features of the environment. If raised in a box, without access to drugs, one never develops a drug problem, regardless of one's propensity to addiction.

In the same way, one breed of dog can be distinguished from others by its propensity to respond to certain events. While all dogs can see birds taking flight in front of them, some are particularly sensitive to the small quick motion of something going aloft. Their threshold to respond to this motion is much lower than for dogs not bred to be hunting companions. By comparison to dogs, our response threshold is higher still. We humans can certainly see the birds taking off, but even when they are directly in front of us, we might not notice them. In hunting dogs, the motion is not only noticed, it is directly connected to another tendency: to pursue prey that moves in just that way. And, of course, one must have birds or birdlike things around for this tendency to lead to bird chasing.

Similarly, a sheepdog who will spend his life herding sheep is one who has a certain set of specific tendencies: to notice and keep track of individuals of a group, to detect the errant motion of a sheep moving away from the herd, and to have a drive to keep the herd together. The end result is a herding dog, but his behavior is made up of piecemeal tendencies that shepherds direct toward controlling their sheep. The dog must also be exposed to sheep early in his life, or these propensities wind up being applied not to sheep, but in a disorganized way to young children, to people jogging in the park, or to the squirrels in your yard.

A dog breed that is called *aggressive,* then, is one that might have a lower threshold to perceive and react to a threatening motion. If the threshold is too low, then even neutral motion—approaching the dog—may be perceived as threatening. But if the dog is not encouraged to follow through on this tendency, it is quite likely that he will never exhibit the aggression that his breed is notorious for.

Knowing the breed of a dog gives us a first-pass entry into understanding something about the dog before we have even met the dog. But it is a mistake to think that knowing a breed guarantees that it will behave as advertised—only that it has certain tendencies. What you get with a mixed-breed dog is a softening of the hard edges seen in breeds. Temperaments are more complex: averaged versions of their bred forebears. In any event, naming a dog's breed is only the beginning of a true understanding of the dog's umwelt, not an endpoint: it doesn't get to what the dog's life is about *to the dog*.

ANIMALS WITH AN ASTERISK

It's snowing and dawn is breaking, which means we have about three minutes for me to get dressed and get us into the park to play before the snow is trammeled by other merrymakers. Outside, well bundled, I plow clumsily through the deep snow, and Pump hurtles herself through it with great bounds, leaving the footprints of a giant bunny. I plop down to make a snow angel, and Pump throws herself down beside me and seems to be making a snow-dog angel, twisting to and fro on her back. I look to her with complete joy at our shared play. Then I smell a horrible odor coming from her direction. The realization is quick: Pump's not making a snow-dog angel; she's rolling in the decaying carcass of a small animal.

There is a tension between those who consider dogs wild animals at their core and those who consider dogs creatures of our own making. The first group tends to turn to wolf behavior to explain dog behavior. The recently popular dog trainers are admired for their full embrace of the wolf side of dogs. They are often seen mocking the second group, which treats their dogs as quadrupedal, slobbery people. Neither has got it right. The answer is plumb in the middle of these approaches. Dogs are animals, of course, with atavistic tendencies, but to stop here is to have a blinkered view of the natural history of the dog. They have been retooled. Now they are animals with an asterisk.

The inclination to look at dogs as animals rather than creations of our psychology is essentially right. To avoid anthropomorphizing, some turn to what might be called unsympathetic biology: a biology free of subjectivity or such messy considerations as consciousness, preferences, sentiment, or personal experiences. A dog is but an animal, they say, and animals are but biological systems whose behavior and physiology can be explained with simpler, general-purpose terminology. Recently I saw a woman leaving the pet store with her terrier, who himself was newly shod in four tiny shoes—to prevent his bringing the street filth into her house, she explained, as she pulled him skating on rigid limbs down the filthy street. This woman could benefit from more reflection on her dog's animal nature, and less on his resemblance to a stuffed toy. In fact, as we'll see, understanding some of the dogs' complexities—the acuity of their noses, what they can see and cannot see, their loss of fearfulness, and the simple affect of a wag—goes a long way to understanding dogs.

On the other hand, in a number of ways, calling a dog *just an animal,* and explaining all dog behavior as emerging from wolf behavior, is incomplete and misleading. The key to dogs' success living with us in our homes is the very fact that dogs are not wolves.

For instance, it is high time we revamp the false notion that our dogs view us as their "pack." The "pack" language—with its talk of the "alpha" dog, dominance, and submission—is one of the most pervasive metaphors for the family of humans and dogs. It originates where dogs originated: dogs emerged from wolflike ancestors, and wolves form packs. Thus, it is claimed, dogs form packs. The seeming naturalness of this move is belied by some of the attributes we *don't* transfer from wolves to dogs: wolves are hunters, but we don't let our dogs hunt for their own food.* And though we may feel secure with a dog at the threshold of a nursery, we would never let a wolf alone in a room with our sleeping newborn baby, seven pounds of vulnerable meat.

Still, to many, the analogy to a dominance-pack organization is terribly appealing—especially with us as dominant and the dog submissive. Once applied, the popular conception of a pack works itself into all sorts of interactions with our dogs: we eat first, the dog second; we command, the dog obeys; we walk the dog, the dog doesn't walk us. Unsure how to deal with an animal in our midst, the "pack" notion gives us a structure.

Unfortunately, it not only limits the kind of understanding and interaction we can have with our dogs, it also relies on a faulty premise. The "pack" evoked in this way bears little resem-

---

*Not only do dogs not typically hunt to feed themselves—whether encouraged to or not—but what hunting technique they have is, it has been noted, "sloppy." A wolf makes a calm, steady track toward his prey, without any frivolous moves; untrained dogs' hunting walks are herky-jerky, meandering back and forth, speeding and slowing. Worse, they may get waylaid by distracting sounds or a sudden urge to playfully pursue a falling leaf. Wolves' tracks reveal their intent. Dogs have lost this intent; we have replaced it with ourselves.

blance to actual wolf packs. The traditional model of the pack was that of a linear hierarchy, with a ruling alpha pair and various "beta" and even "gamma" or "omega" wolves below them, but contemporary wolf biologists find this model far too simplistic. It was formed from observations of *captive* wolves. With limited space and resources in small, enclosed pens, unrelated wolves self-organize, and a hierarchy of power results. The same might happen in any social species confined with little room.

In the wild, wolf packs consist almost entirely of related or mated animals. They are *families,* not groups of peers vying for the top spot. A typical pack includes a breeding pair and one or many generations of their offspring. The pack unit organizes social behavior and hunting behavior. Only one pair mates, while other adult or adolescent pack members participate in raising the pups. Different individuals hunt and share food; at times, many members together hunt large prey which may be too large to tackle individually. Unrelated animals do occasionally join together to form packs with multiple breeding partners, but this is an exception, probably an accommodation to environmental pressures. Some wolves never join a pack.

The one breeding pair—parents to all or most of the other pack members—guides the group's course and behaviors, but to call them "alphas" implies a vying for the top that is not quite accurate. They are not alpha dominants any more than a human parent is the alpha in the family. Similarly, the subordinate status of a young wolf has more to do with his age than with a strictly enforced hierarchy. Behaviors seen as "dominant" or "submissive" are used not in a scramble for power, they are used to maintain social unity. Rather than being a pecking order, rank is a mark of age. It is regularly on display in the animals' expressive postures in greeting and in interaction. Approaching an older wolf with a low wagging tail and a body close to the ground, a younger wolf is acknowledging the older's biological priority. Young pups are naturally at a subordinate level; in mixed-family packs pups may inherit some of the status of their parents. While rank may be reinforced by charged and sometimes dangerous encounters between pack members, this is rarer than aggression against an intruder. Pups learn their place by interacting with and observing their packmates more than by being put in their place.

The reality of wolf pack behavior contrasts starkly with dog behavior in other ways. Domestic dogs do not generally hunt. Most are not born into the family unit in which they will live: with humans the predominant members. Pet dogs' attempts to mate are (happily) unrelated to their adopted humans'—supposedly the alpha pair's—mating schedules. Even feral dogs—those who may never have lived in a human family—usually do not form traditional social packs, although they may travel in parallel.

Neither are we the dog's pack. Our lives are so much more stable than that of a wolf pack: the size and membership of a wolf pack is always in flux, changing with the seasons, with the rates of offspring, with young adult wolves growing up and leaving in their first years, with the availability of prey. Typically, dogs adopted by humans live out their lives with us; no one is pushed out of the house in spring or joins us just for the big winter moose hunt. What domestic dogs do seem to have inherited from wolves is the sociality of a pack: an interest in being around others. Indeed, dogs are social opportunists. They are attuned to the actions of others, and humans turned out to be very good animals to attune to.

To evoke the outdated, simplistic model of packs glosses over real differences between dog and wolf behavior and misses some of the most interesting features of packs in wolves. We do better to explain dogs' taking commands from us, deferring to us, and indulging us by the fact that we are their source of food than by reasoning that we are alpha. We can certainly make dogs totally submissive to us, but that is neither biologically necessary nor particularly enriching for either of us. The pack analogy does nothing but replace our anthropomorphisms with a kind of "beastomorphism," whose crazy philosophy seems to be something like "dogs aren't humans, so we must see them as precisely un-human in every way."

We and our dogs come closer to being a benign gang than a pack: a gang of two (or three or four or more). We are a family. We share habits, preferences, homes; we sleep together and rise together; we walk the same routes and stop to greet the same dogs. If we are a gang, we are a merrily navel-gazing gang, worshiping nothing but the maintenance of our gang itself. Our gang works by sharing fundamental premises of behavior. For instance, we agree to rules of conduct in our home. I agree with my family that under no circumstances is urination on the living room rug allowed. This is a tacit agreement, happily. A dog has to be taught this premise for habitation; no dog knows about the value of rugs. In fact, rugs might provide a nice feeling underfoot for some bladder release.

Trainers who espouse the pack metaphor extract the "hierarchy" component and ignore the social context from which it emerges. (They further ignore that we still have a lot to learn about wolf behavior in the wild, given the difficulty of following these animals closely.) A wolfcentric trainer may call the humans the pack leaders responsible for discipline and forcing submission by others. These trainers teach by punishing the dog after discovery of, say, the inevitable peed-upon rug. The punishment can be a yell, forcing the dog down, a sharp word or jerk of the collar. Bringing the dog to the scene of the crime to enact the punishment is common—and is an especially misguided tactic.

This approach is farther from what we know of the reality of wolf packs and closer to the timeworn fiction of the animal kingdom with humans at the pinnacle, exerting dominion over the rest. Wolves seem to learn from each other not by punishing each other but by observing each other. Dogs, too, are keen observers—of our reactions. Instead of a punishment happening *to* them, they'll learn best if you let them discover for themselves which behaviors are rewarded and which lead to naught. Your relationship with your dog is defined by what happens in those undesired moments—as when you return home to a puddle of urine on the floor. Punishing the dog for his misbehavior—the deed having been done maybe hours before—with dominance tactics is a quick way to make your relationship about bullying. If your trainer punishes the dog, the problem behavior may temporarily abate, but the only relationship created is one between your trainer and your dog. (Unless the trainer's moving in with you, that won't last long.) The result will be a dog who becomes extra sensitive and possibly fearful, but not one who understands what you mean to impart. Instead, let the dog use his observation skills. Undesired behavior gets no attention, no food: nothing that the dog wants from you. Good behavior gets it all. That's an integral part of how a young child learns how to be a person. And that's how the dog-human gang coheres into a family.

CANIS UNFAMILIARIS

On the other hand, let's not forget that it is only tens of thousands of years of evolution that separate wolves and dogs. We would have to go back millions of years to trace our split from chimpanzees; appropriately, we do not look to chimpanzee behavior to learn how to raise our children.* Wolves and dogs share all but a third of 1 percent of their DNA. We see occasional snatches of wolf in our pets: a glimpse of a growl when you move to extract a beloved ball from your dog's mouth; rough-and-tumble play in which one animal seems more prey than playmate; a glimmer of wildness in the eye of a dog grabbing for a meat bone.

The orderliness of most of our interactions with dogs clashes mightily with their atavistic side. Once in a while it feels as if some renegade ancient gene takes a hold of the domesticated product of its peers. A dog bites his owner, kills the family cat, attacks a neighbor. This unpredictable, wild side of dogs should be acknowledged. The species has been bred for millennia, but it evolved for millions of years before that without us. They were predators. Their jaws are strong, their teeth designed for tearing flesh. They are wired to act before contemplating action. They have an urge to protect—themselves, their families, their turf—and we cannot always predict when they will be prompted to be protective. And they do not automatically heed the shared premises of humans living in civilized society.

As a result, the first time your dog tears from your side, running maniacally off the trail after some invisible thing in the bushes, you panic. With time, you will become familiar with each other: they, with what you expect of them; you, with what they do. It is only *off the trail* to you; to the dog it is a natural continuation of walking, and he will learn about trails in time. You may never see the invisible thing in the bush, but you learn, after

---

*Notably, the number of behavioral similarities between chimpanzees and humans (culture and language aside for the moment) increases steadily as the number of scientific studies of the chimps also steadily increases.

a dozen walks, that invisible things are in bushes, and the dog will return to you. Living with a dog is a long process of becoming mutually familiar. Even the dog bite is not a uniform entity. There are bites done out of fear, out of frustration, out of pain, and out of anxiety. An aggressive snap is different than an exploratory mouthing; a play bite is different than a grooming nibble.

Despite their sometime wildness, dogs never revert to wolves. Stray dogs—those who lived with humans but have wandered away or been abandoned—and free-ranging dogs—provisioned with food but living apart from humans—do not take on more wolflike qualities. Strays seem to live a life familiar to city dwellers: parallel to and cooperative with others, but often solitary. They do not self-organize socially into packs with a single breeding pair. They don't build dens for the pups or provide food for them as wolves do. Free-ranging dogs may form a social ordering like other wild canids—but one organized by age more than by fights and strife. Neither hunts cooperatively: they scavenge or hunt small prey by themselves. Domestication changed them.

Even when wolves have been socialized— raised from birth among humans instead of other wolves—they do not turn into dogs. They strike a middle ground in behavior. Socialized wolves are more interested in and attentive to humans than wild-born wolves. They follow human communicative gestures better than wild wolves. But they are not dogs in wolves' clothing. Dogs raised with a human caretaker prefer her company over that of other humans; wolves are less discriminating. Dogs far outpace hand-raised wolves in interpreting human cues. To see a wolf on a leash, sitting and lying down on request, one could be convinced there is little difference between the socialized wolf and dog. To see that wolf in the presence of a rabbit is to see how much difference there still is: the human is forgotten while the rabbit is relentlessly pursued. A dog near that same rabbit may patiently wait, gazing at his owner, to be permitted to run. Human companionship has become dogs' motivational meat.

MAKING YOUR DOG

As you choose a new dog from among a litter or a loud shelter of baying mutts and bring him home, you begin to "make a dog" again, recapitulating the history of domestication of the species. With each interaction, with each day, you define—at once circumscribing and expanding—his world. In the first few weeks with you, the pup's world is, if not entirely a tabula rasa, awfully close to the "blooming, buzzing confusion" that a newborn baby experiences. No dog knows, on first turning his eyes on the person who peeks at him in his shelter cage, what the person expects of him. Many people's expectations, at least in this country, are fairly similar: be friendly, loyal, pettable; find me charming and lovable—but know that I am in charge; do not pee in the house; do not jump on guests; do not chew my dress shoes; do not get into the trash. Somehow, word hasn't gotten to the dogs. Each dog has to be taught this set of parameters for his life with people. The dog learns, through you, the kinds of things that are important to you—and that you want to be important to him. We are all domesticated, too: inculcated with our culture's mores, with how to be human, with how to behave with others. This is facilitated by language, but spoken language is not necessary to achieve it. Instead we need to be alert to what the dog is perceiving and to make our perceptions clear to him.

The first-century Roman encyclopedist Pliny's prodigious *Natural History* includes a confident statement of fact about the birth of bears. The cubs, he wrote, "are a white and shapeless lump of flesh, little larger than mice, without eyes or hair and only the claws projecting. This lump the mother bears slowly lick into shape." The bear is born, he was suggesting, as nothing but pure undifferentiated matter, and, like a true empiricist, the mother bear *makes* her pup a bear by licking it. When we brought Pump into our home, I felt I was doing just this: I was licking her into shape. (And not just because there was a lot of licking between us—after all, it was exclusively she who was licking.) It was our way of interacting together that made her who she was, that makes dogs that most people want to live with: interested in our goings and comings, attentive to us, not overly intrusive, playful just at the right times. She interpreted the world through acting on it, by seeing others act, by being shown, and by acting with me on the world—promoted into being a good member of the family. And the more time we spent together, the more she became who she was, and the more we were intertwined.

Sniff

First sniff of the day: Pump wanders into the living room in the morning while I am dishing out her food. She's looking sleepy but her nose is wide awake, stretching every which way as though it's doing morning exercises. She reaches her nose toward the food without committing her body, and sniffs. A look at me. Another sniff. A judgment has been levied. She backs from the bowl and forgives me by nosing my outstretched hand, her whiskers tickling while her moist nose examines my palm. We go outside and her nose is gymnastic, almost prehensile, happily taking in smells that gust by . . .

We humans tend not to spend a lot of time thinking about smelling. Smells are minor blips in our sensory day compared to the reams of visual information that we take in and obsess over in every moment. The room I'm in right now is a phantasmagoric mix of colors and surfaces and densities, of small movements and shadows and lights. Oh, and if I really call my attention to it I can smell the coffee on the table next to me, and maybe the fresh scent of the book cracked open—but only if I dig my nose into its pages.

Not only are we not always smelling, but when we do notice a smell it is usually because it is a good smell, or a bad one: it's rarely just a source of information. We find most odors either alluring or repulsive; few have the neutral character that visual perceptions do. We savor or avoid them. My current world seems relatively odorless. But it is most decidedly not free of smell. Our own weak olfactory sense has, no doubt, limited our curiosity about what the world smells like. A growing coalition of scientists is working to change that, and what they have found about olfactory animals, dogs included, is enough to make us envy those nose-creatures. As we *see* the world, the dog *smells* it. The dog's universe is a stratum of complex odors. The world of scents is at least as rich as the world of sight.

SNIFFERS

. . . Her ungulate-grazing sniff, nose deep in a patch of good grass, trawling the ground and not coming up for air; the examinatory sniff, judging a proffered hand; the alarm-clock sniff, close enough to my sleeping face to tickle me awake with her whiskers; the contemplative sniff, nose held high in the wake of a breeze. All followed by a half sneeze—just the CHOO, no AH—as though to clear her nostrils of whatever molecule she'd just inhaled . . .

Dogs don't act on the world by handling objects or by eyeballing them, as people might, or by pointing and asking others to act on the object (as the timid might); instead they bravely stride right up to a new, unknown object, stretch their magnificent snouts within millimeters of it, and take a nice deep sniff. That dog nose, in most breeds, is anything but subtle. The snout holding the nose projects forth to examine a new person seconds before the dog himself arrives on the scene. And the sniffer is not just an ornament atop the muzzle; it is the leading, moist headliner. What its prominence suggests, and what all science confirms, is that the dog is a creature of the nose.

The sniff is the great medium for getting smelly objects to the dog, the tramway on which chemical odors speed up to the waiting receptor cells along the caverns of the dog nose. Sniffing is the action of inhaling air, but it is more active than that, usually involving short, sharp bursts of drawing air into the nose. Everyone sniffs—to clear the nose, to smell dinner cooking, as part of a preparatory inhale. Humans even sniff emotively, or meaningfully—to express disdain, contempt, surprise, and as punctuation at a sentence's end. Animals mostly sniff, as far as we know, to investigate the world. Elephants raise their trunk into the air in a "periscope sniff," tortoises slowly reach and open their nostrils wide in a sniff, marmosets sniff while they nuzzle. Ethologists watching animals often take note of all these sniffs, for they may precede an attempt to mate, a social interaction, aggression, or feeding. They record an animal as "sniffing" when it brings its nose close to—but not touching—the ground or an object, or an object is brought close to—but not touching—the nose. In these cases, they are assuming that the animal is in fact inhaling sharply—but they may not be able to get close enough to the animal to see the nostrils moving, or the tiny vortex of air that stirs the area in front of the nose.

Few have looked closely at exactly what happens in a sniff. But recently some researchers have used a specialized photographic method that shows air flow in order to detect when, and how, dogs are sniffing. They have found that the sniff is nothing to be sniffed at. In fact one could make the case that it is neither a single nor a simple inhalation. The sniff begins with muscles in the nostrils straining to draw a current of air into them—this allows a large amount of any air-based odorant to enter the nose. At the same time, the air already in the nose has to be displaced. Again, the nostrils quiver slightly to push the present air deeper into the nose, or off through slits in the side of the nose and backward, out the nose and out of the way. In this way, inhaled odors don't need to jostle with the air already in the nose for access to the lining of the nose. Here's why this is particularly special: the photography also reveals that the slight wind generated by the exhale in fact helps to pull more of the new scent in, by creating a current of air over it.

This action is markedly different from human sniffing, with our clumsy "in through one nostril hole, out through the same hole" method. If we want to get a good smell of something, we have to sniff-hyperventilate, inhaling repeatedly without strongly exhaling. Dogs naturally create tiny wind currents in exhalations that hurry the inhalations in. So for dogs, the sniff includes an exhaled component that helps the sniffer smell. This is visible: watch for a small puff of dust rising up from the ground as a dog investigates it with his nose.

Given our tendency to find so many smells disgusting, we should all celebrate that our olfactory system adapts to an odor in the environment: over time, if we stay in one place, the intensity of every smell diminishes until we don't notice it at all. The first smell of coffee brewing in the morning: fantastic . . . and gone in a few minutes. The first smell of something rotting under the porch: nauseating . . . and gone in a few minutes. The sniffing method of dogs enables them to avoid habituation to the olfactory topography of the world: they are continually refreshing the scent in their nose, as though shifting their gaze to get another look.

THE NOSE NOSE

> I crack open her window in the car—just enough to fit a dog-sized head (remembering the time she threw herself completely out the open window after that squirrel hitchhiking on the side of the road). Pump props herself up on the armrest and pokes her muzzle out of the car as we race along in the night. She squints her eyes tight, her face is streamlined in the wind, and she projects her nose deep into the rushing air.

Once a smell has been vacuumed in, it finds a receptive welcome from an extravagance of nasal tissue. Most purebreds, and nearly all mutts, have long muzzles in whose noses are labyrinths of channels lined with special skin tissue. This lining, like the lining of our own noses, is primed to receive air carrying "chemicals"—molecules of various sizes that will be perceived as scents. Any object we encounter in the world is cast in a haze of these molecules—not only the ripe peach on the counter but the shoes we kick off at the door and the doorknob we grasp. The tissue of the inside of the nose is entirely blanketed with tiny receptor sites, each with soldiers of hairs to help catch molecules of certain shapes and pin them down. Human noses have about six million of these sensory receptor sites; sheepdog noses, over two hundred million; beagle noses, over three hundred million. Dogs have more genes committed to coding olfactory cells, more cells, and more *kinds* of cells, able to detect more kinds of smells. The difference in the smell experience is exponential: on detecting certain molecules from that doorknob, not single sites but combinations of sites fire together to send information to the brain. Only when the signal reaches the brain is it experienced as a scent: if it is we doing the sniffing, we'd say *A-ha! I smell it.*

More than likely, though, we won't smell it. But the beagle will: it's been estimated that their sense of smell may be millions of times more sensitive than ours. Next to them we are downright anosmic: smelling nothing. We might notice if our coffee's been sweetened with a teaspoon of sugar; a dog can detect a teaspoon of sugar diluted in a million gallons of water: two Olympic-sized pools full.*

What's this like? Imagine if each detail of our visual world were matched by a corresponding smell. Each petal on a rose may be distinct, having been visited by insects leaving pollen footprints from faraway flowers. What is to us just a single stem actually holds a record of who held it, and when. A burst of chemicals marks where a leaf was torn. The flesh of the petals, plump with moisture compared to that of the leaf, holds a different odor besides. The fold of a leaf has a smell; so does a dewdrop on a thorn. And *time* is in those details: while we can see one of the petals drying and browning, the dog can smell this process of decay and aging. Imagine smelling every minute visual detail. That might be the experience of a rose to a dog.

The nose is also the fastest route by which information can get to the brain. While visual or auditory data goes through an intermediate staging ground on the way to the cortex, the highest level of processing, the receptors in the nose connect directly to nerves in specialized olfactory "bulbs" (so shaped). The olfactory bulbs of the dog brain make up about an eighth of its mass:

*. . . theoretically: no swimming pools have been used in such a test. Instead, experimenters use extremely small samples of an odorless medium, and then add an even more extremely small sample of sugar to one of them.

proportionally greater than the size of our central visual processing center, the occipital lobes, in our brains. But dogs' specially keen sense of smell may also be due to an additional way they perceive odors: through the vomeronasal organ.

THE VOMERONASAL NOSE

What specificity of image the name "vomeronasal" conjures up! Evoking the displeasure of getting a good sniff of fresh vomit, the "vomer" is actually a description of the part of the small bone in the nose where the sensory cells sit. Still, the name seems somehow fitting for an animal that is notorious for coprophagia (feces eating) and that may lick another dog's urine off the ground. Neither act is vomitous for dogs; it's just a way of getting even more information about other dogs or animals in the area. The vomeronasal organ, first discovered in reptiles, is a specialized sac above the mouth or in the nose covered with more receptor sites for molecules. Reptiles use it to find their way, to find food, and to find mates. The lizard who darts out its tongue to touch an unknown object is not tasting or sniffing; it is drawing chemical information toward its vomeronasal organ.

These chemicals are pheromones: hormonelike substances released by one animal and perceived by another of the same species, and usually prompting a specific reaction—such as readying oneself for sex—or even changing hormonal levels. There is some evidence that humans unconsciously perceive pheromones, perhaps even through a nasal vomeronasal organ.*

---

*The psychologist Martha McClintock was the first to seriously study pheromone detection in humans; she and others have done savvy, fascinating studies of how our behavior and hormonal rates may be affected by pheromones or pheromone-like hormones. But the jury is still out—and loudly arguing—on these claims.

Dogs definitely have a vomeronasal organ: it sits above the roof (hard palate) of the mouth, along the floor of the nose (nasal septum). Unlike in other animals, the receptor sites are covered in cilia, tiny hairs encouraging these molecules along. Pheromones are often carried in a fluid: urine, in particular, is a great medium for one animal to send personalized information to members of the opposite sex about, say, one's eagerness to mate. To detect the pheromones in that urine some mammals touch the liquid and do a distinctive, mortifying, lip-curling grimace called *flehmen*. The face of a flehmening animal is notably unlovable—but it is the face of an animal who is on the hunt for a lover. The flehmen pose seems to hurtle the fluid toward the animal's vomeronasal organ, where it is pumped into the tissue, or is absorbed through capillary action. Rhinos, elephants, and other ungulates flehmen regularly; so do bats and cats, which have their own species variations. Humans may have vomeronasal organs, but we do not flehmen. Neither do dogs. But a regular observer of dogs will notice an often very intense interest in the urine of other dogs—sometimes an interest which lures them right . . . up . . . into . . . wait, gross! Stop licking that! Dogs may lightly lap up urine, especially urine of a female in heat. This could be their version of flehmen.

Even better than flehmen is keeping the outside of the nose nice and moist. The vomeronasal organ is probably why a dog's nose is wet. Most animals with vomeronasal organs have wet noses, too. It is difficult for an airborne odor to land squarely on the vomeronasal organ, since it is situated in a safe, dark interior recess of the face. A hearty sniff not only brings molecules into the dog's nasal cavity; little molecular bits also stick onto the moist exterior tissue of the nose. Once there, they can dissolve and travel to the vomeronasal organ through interior ducts. When your dog nuzzles against you, he is actually collecting your odor on his nose: better to confirm that you're you. In this way, dogs double their methods of smelling the world.

THE BRAVE SMELL OF A STONE

When Pump got her nose into a good smell in the grass—when she really dug her nose deeply into the earth—I came to know what was going to happen next. She'd hop around, resniff the smell from different angles, then take a tentative swipe at it, upending a dollop of turf. More deep sniffing, some licking, smushing her nose into the ground—and then the climax: an unrestrained dive into the smell, nose first, throwing her whole body down after it, and wriggling madly back and forth.

What, then, do these noses enable the dog to smell? What does the world look like from the vantage of a nose? Let's start with the easy stuff for them: what they smell of us and of each other. Then we might be ready to challenge them to smell time, the history of a river stone, and the approach of a thunderstorm.

The smelly ape

Humans stink. The human armpit is one of the most profound sources of odor produced by any animal; our breath is a confusing melody of smells; our genitals reek. The organ that covers our body—our skin—is itself covered in sweat and sebaceous glands, which are regularly churning out fluid and oils holding our particular brand of scent. When we touch objects, we leave a bit of ourselves on them: a slough of skin, with its clutch of bacteria steadily munching and excreting away. This is our smell, our signature odor. If the object is porous—a soft slipper, say— and we spend a lot of time touching it—putting a foot in it, clutching it, carrying it under an arm—it becomes an extension of ourselves for a creature of the nose. For my dog, my slipper is a part of me. The slipper may not look to us like an object that would be terribly interesting to a dog, but anyone who has returned home to find a ravaged slipper, or who has been tracked by the scent they've left thereon, knows otherwise.

We needn't even touch objects for them to smell of us: as we move, we leave behind a trail of skin cells. The air is perfumed with our constant dehumidifying sweat. Added to this, we wear in odor what we've eaten today, whom we've kissed, what we've brushed against. Whatever cologne we put on merely adds to the cacophony. On top of this, our urine, traveling down from the kidneys, catches odorous notes from other organs and glands: the adrenal glands, the renal tubes, and potentially the sex organs. The trace of this concoction on our bodies and our clothes provides more uniquely specific information about us. As a result, dogs find it incredibly easy to distinguish us by scent alone. Trained dogs can tell identical twins apart by scent. And our aroma remains even when we've left, hence the "magical" powers of tracking dogs. These skilled sniffers see us in the cloud of molecules we leave behind.

To dogs, we *are* our scent. In some ways, olfactory recognition of people is quite similar to our own visual recognition of people: there are multiple components of the image responsible for how we look. A different haircut or a newly bespectacled face can, at least momentarily, mislead us as to the identity of the person standing before us. I can be surprised at what even a close friend looks like from a different vantage or from a distance. So too must the olfactory image we embody be different in different contexts. The mere arrival of my (human) friend at the dog park is enough to set me smiling; it takes another beat before my dog notices her own friend. And odors are subject to decay and dispersal that light is not: a smell from a nearby object may not reach you if a breeze carries it in the other direction, and the strength of an odor diminishes over time. Unless my friend tries ducking behind a tree, it's hard for her to conceal her visual image from me: a wind won't conceal her. But it might conceal her from a dog momentarily.

When we return home at a day's end, dogs typically greet our cocktail of stink promptly and lovingly. Should we come home after having bathed in unfamiliar perfume or wearing someone else's clothes, we might expect a moment of puzzlement—it is no longer "us"—but our natural effusion will soon give us away. Dogs are not alone among animals in seeing in scent. Sharks have been seen to follow the same zigzaggy path through water that an injured fish took some time before: through not just its blood but also its hormones, the fish has left a bit of itself behind. But dogs are unique in being encouraged and trained by people to use scent to follow someone who is visually long gone.

Bloodhounds are one of the supersmellers among dogs. Not only do they have more nose tissue—more *nose*—but many features of their body seem to conspire to enable them to smell extra strongly. Their ears are terrifically long, but not to enable better hearing, as they fall close to the head. Instead a slight swing of the head sets these ears in motion, fanning up more scented air for the nose to catch. Their constant stream of drool is a perfect design to gather extra liquids up to the vomeronasal organ for examination. Basset hounds, thought to be bred from bloodhounds, go one step further: with their foreshortened legs, the whole head is already at ground—scent—level.

These hounds smell well naturally. Through training— rewarding them for attending to certain scents and ignoring others—they are easily able to follow a scent left by someone one or many days before, and can even specify where two individuals parted ways. It doesn't take very much of our odor: some researchers tested dogs using five thoroughly cleaned glass slides, to one of which a single fingerprint was added. The slides were put away for a few hours or up to three weeks. Dogs then got to examine the array of slides, and tried to choose the human slide: they were rewarded with a treat if they guessed correctly, which is sufficient motivation for them to stand and sniff at glass slides. One dog was correct on all but six of one hundred trials. When the slides were then placed outside on the building roof for a week, exposed over the course of the seven days to direct sun, rain, and all manner of blowing debris, the same dog was still correct on almost half the trials—well above chance.

They track not just by noticing odors, but by noticing very small changes in odor. Each of our footsteps will have more or less the same amount of our scent in it. In theory, then, if I saturate the ground with my scent, by running chaotically to and fro, a dog who tracks by noticing that smell won't be able to tell my path—only that I've definitely been there. But trained dogs don't just notice a smell. They notice the change in a smell over time. The concentration of an odor left on the ground by, say, a running footprint, diminishes with every second that passes. In just two seconds, a runner may have made four or five footprints: enough for a trained tracker to tell the direction that he ran based just on the differences in the odor emanating from the first and fifth print. The track you left as you exited the room has more smell in it than the one right before it; thus your path is reconstructed. Scent marks time.

Conveniently, instead of becoming inured to smells over time, as we do, the vomeronasal organ and the dog nose may regularly swap roles, to keep the scent fresh. It is this ability that is exploited when training rescue dogs, who must orient themselves to the odor of someone who has disappeared. Similarly, scenting dogs who trail a criminal suspect are trained to follow what is delicately called our "personal odor generation": our natural, regular, and entirely involuntary butyric acid production. This is easy for them, and they can then extend this skill to smelling other fatty acids, too. Unless you are wearing a body suit made entirely of scentproof plastic, a hound can find you.

You showed fear

Even those of us who are not fleeing a crime scene, or in need of rescue, have a reason not to underestimate just how good a sniffer the dog is. Not only can dogs identify individuals by odor, they can also identify *characteristics* of the individual. A dog knows if you've had sex, smoked a cigarette (done both these things in succession), just had a snack, or just run a mile. This may seem benign: except, perhaps, for the snack, these facts about you might not be of particular interest to a dog. But they can also smell your emotions.

Generations of schoolchildren have been admonished to "never show fear" to a strange dog.* It is likely that dogs do smell fear, as well as anxiety and sadness. Mystical abilities need not be

---

*This construction—*strange dog*—itself seems geared to inspire fear. Its use is also based on a flawed premise: that familiar dogs will behave predictably and reliably, and unfamiliar ones will not. As we've seen, as much as we may want dogs to behave in lockstep with our desires, their simply being their own animals ensures that they will not always do so.

invoked to account for this: fear *smells*. Researchers have identified many social animals, from bees to deer, who can detect pheromones emitted when one animal is alarmed, and who react by taking action to get to safety. Pheromones are produced involuntarily and unconsciously, and through different means: damaged skin may provoke release of them, and there are specialized glands that release chemicals of alarm. In addition, the very feeling of alarm, fear, and every other emotion correlates with physiological changes, from changes in heart rate and breathing rate, to sweating and metabolic changes. Polygraph machines work (to the extent that they work at all) by measuring changes in these autonomic bodily responses; one might say that animals' noses "work" by being sensitive to them as well. Laboratory experiments using rats confirm this: when one rat is given a shock in a cage, and learns to be fearful of the cage, other rats nearby pick up on the shocked rat's fear—even without seeing the rat being shocked—and themselves avoid the cage, which was otherwise not distinguishable from nearby cages.

How does that strange, menacing-looking dog smell our apprehension or fear as he approaches us? We spontaneously sweat under stress, and our perspiration carries a note of our odor on it: that's the first clue to the dog. Adrenaline, used by the body to gear up for a good sprint away from something dangerous, is unscented to us, but not to the sensitive sniffer of the dog: another hint. Even the simple act of increased blood flow brings chemicals more quickly to the surface of the body, where they can be diffused through the skin. Given that we emit odors that reflect these physiological changes accompanying fear, and given the budding evidence of pheromones in humans, chances are that if we've got the heebie-jeebies, a dog can tell. And as we'll see later, dogs are skilled readers of our behavior. We can sometimes see fear in other people in their facial expressions;

there is sufficient information in our posture and gait for a dog to see it, too.

In these ways, the fleeing criminal being tracked by dogs is doubly doomed. Dogs can be trained to track based not just on pursuit of a specific person's odor, but also based on a certain kind of odor: the most recent odor of a human in the vicinity (good for finding someone's hiding place), or a human in emotional distress—fearful (as one running from the cops might be), angry, even annoyed.

The smell of disease

If dogs can detect trace amounts of chemicals we leave behind on a doorknob, or in a footprint, might they be able to detect chemicals indicating disease? If you're lucky, when you come down with a disease difficult to diagnose, you'll have a doctor who recognizes, as some have, that a distinctive smell of freshly baked bread about you is due to typhoid fever, or that a stale, sour scent is due to tuberculosis being exhaled from your lungs. According to many doctors, they have come to notice a distinctive smell to various infections, or even to diabetes, cancer, or schizophrenia. These experts come unequipped with the dog's nose—but more equipped to identify disease. Still, a few small-scale experiments indicate that you might get an even more refined diagnosis if you make an appointment with a well-trained dog.

Researchers have begun training dogs to recognize the chemical smells produced by cancerous, unhealthy tissues. The training is simple: the dogs were rewarded when they sat or lay down next to the smells; they weren't rewarded when they didn't. Then the scientists collected the smells of cancer patients and patients without cancer, in small urine samples or by having them breathe into tubes able to catch exhaled molecules. Although the numbers of trained dogs are small, the results were big: the dogs could detect which of the patients had cancer. In one study, they only missed on 14 out of 1,272 attempts. In another small study with two dogs, they sniffed out a melanoma nearly every time. The latest studies show trained dogs can detect cancers of the skin, breast, bladder, and lungs at high rates.

Does this mean your dog will let you know when a small tumor develops in you? Probably not. What it indicates is that dogs are *able* to do so. You might smell different to them, but your changing smell might be gradual. Both you and your dog would need training: the dog to pay attention to the smell, you to pay attention to behaviors indicating your dog has found something.*

The smell of a dog

Since odor is so conspicuous to a dog, it gets great use socially. While we humans leave our scents behind inadvertently, dogs are not only advertent, they are profligate with their scents. It is as though dogs, realizing how well the odor of our bodies comes to stand for ourselves (even in our absence), determined to use this to their advantage. All canids—wild and domestic dogs

---

*Research on other diseases is proceeding apace. Provocatively, dogs who live in homes with epileptics seem to be moderately good predictors of seizures. Two studies report that dogs licked the person's face or hands, whimpered, stood nearby, or moved protectively—in one case sitting *on* a child, and in another blocking a child's access to stairs—before seizures. If this is true, there may be olfactory, visual, or some other invisible (to us) cues that the dogs used. But as the data come from "self-report"—family questionnaires rather than data gathered more objectively—more evidence is needed. We can however, pause in admiration of the possibility of such a skill.

and their relations—leave urine conspicuously splashed on all manner of object. Urine marking, as this method of communication is called, conveys a message—but it is more like note-leaving than a conversation. The message is left by one dog's rear end for retrieval by another's front end. Every dog owner is familiar with the raised-leg marking of fire hydrants, lampposts, trees, bushes, and sometimes the unlucky dog or bystander's pant leg. Most marked spots are high or prominent: better to be seen, and better for the odor in the urine (the pheromones and affiliated chemical stew) to be smelled. Dogs' bladders—sacs that serve no known purpose except as a holding pen for urine—allow for release of just a little urine at a time, allowing them to mark repeatedly and often.

And having left smells in their wake, they also come right up to investigate others' smells. From observations of the behavior of the sniffing dogs, it appears that the chemicals in the urine give information about, for females, sexual readiness, and for males, their social confidence. The prevailing myth is that the message is "this is mine": that dogs urinate to "mark territory." This idea was introduced by the great early-twentieth-century ethologist Konrad Lorenz. He formed a reasonable hypothesis: urine is the dog's colonial flag, planted where one claims ownership. But research in the fifty years since he proposed that theory has failed to bear that out as the exclusive, or even predominant, use of urine marking.

Research on free-ranging dogs in India, for instance, showed how dogs behave when left entirely to their own devices. Both sexes marked, but only 20 percent of the markings were "territorial"—on a boundary of a territory. Marking changed by seasons, and happened more often when courting or when scavenging. The "territory" notion is also belied by the simple fact that few dogs urinate around the interior corners of the house or apartment where they live. Instead, marking seems to leave information about who the urinator is, how often he walks by this spot in the neighborhood, his recent victories, and his interest in mating. In this way, the invisible pile of scents on the hydrant becomes a community center bulletin board, with old, deteriorating announcements and requests peeking out from underneath more recent posts of activities and successes. Those who visit more frequently wind up being at the top of the heap: a natural hierarchy is thus revealed. But the old messages still get read, and they still have information—one element of which is simply their age.

In the annals of animal urine marking, dogs are not the most impressive players. Hippopotami wave their tails as they spray urine, better to scatter it, sprinklerlike, in all directions. There are rhinoceroses who follow their high-powered urination onto bushes with destruction of the same bushes with horn and hoof—to ensure, presumably, that their urine is spread far and wide. Pity the owner whose dog is the first to discover the spreading-efficiency of high-powered, whirling-sprinkler urination.

Other animals also press their rear ends against the ground to release fecal and other anal odors. The mongoose does a handstand and rubs itself against a high perch; some dogs do what gymnastics they can, seemingly deliberately relieving themselves on large rocks and other outcroppings. Although secondary to urine marking, defecation also holds identifying odors—not in the excreta itself but in the chemicals dolloped on top. These come from the pea-sized anal sacs, situated right inside the anus and holding secretions from nearby glands: extremely foul, dead-fish-in-a-sweatsock kind of secretions with apparently individual-dead-fish-in-individual-sweatsock odors for each individual dog. These anal sacs also release involuntarily when a dog is afraid or alarmed. It may be no wonder that so many dogs fright at their veterinarian's office: as part of a routine examination, vets often express (squeeze to release the contents of) the anal sacs, which can get impacted and infected. The smell, covered for us by the familiar scent of veterinary antibiotic soaps, must be all over the vets: they reek of epic dog fear.

Finally, if these mephitic calling cards are insufficient, dogs have one other trick in their marking book: they scratch the ground after defecation or urination. Researchers think that this adds new odors to the mix—from the glands on the pads of the feet—but it may also serve as a complementary visual cue leading a dog to the source of the odor for closer examination. On a windy day, dogs may seem friskier, more likely to scratch the ground; they may in fact be leading others to a message that otherwise would waft away.

LEAVES AND GRASS

Science, out of decorum or disinterest, has not definitively explained Pump's mad wriggling in a funky spot of grass. The odor may be of a dog she's interested in, or of a dog she recognizes. Or it may be the remnants of a dead animal, rolled in not so much to conceal her own smell as enjoyed for its sumptuous bouquet.

We respond pithily and with soap: by giving our dogs frequent baths. My neighborhood has not only its fill of dog groomers, but is visited by a mobile grooming van that will come to your home to pick up, suds, fluff, and otherwise de-dog your dog for you. I'm sympathetic to owners who have a lower tolerance for detritus and dust around their homes than I do: a well-walked, thoroughly played-out dog is an efficient spreader of dirt. But we deprive our dogs of something by bathing them so much—to say nothing of our culture's overenthusiastic clean-ing of our own homes, including our dogs' bedding. What smells clean to us is the smell of artificial chemical clean, some-thing expressly non-biological. The mildest fragrance that cleansers come in is still an olfactory insult to a dog. And although we might like a visually clean space, a place rid entirely of organic smells would be an impoverished one for dogs. Better to keep the occasional well-worn T-shirt around and not scrub the floors for a while. The dog himself does not have any drive to be what we would call clean. It is no wonder that the dog follows his bath by hightailing it to roll vigorously on the rug or in the grass. We deprive dogs of an important part of their identity, tem-porarily, to bathe them in coconut-lavender shampoo.

Similarly, recent research found that when we give dogs antibiotics excessively, their body odor changes, temporarily wreaking havoc with the social information they normally emit. We can be alert to this while still using these medicines appropri-ately. So too with the laughable Elizabethan collar, an enor-mous cone collar typically used to prevent a dog from chewing at stitches closing a wound: it is useful to prevent self-mutilation, but consider all the ordinary interactive behavior it prevents— looking away from an aggressive dog; seeing someone's loping approach from the side; the ability to reach and sniff another dog's rump.

Pity the urban dog, subjected to the remnants of an old soci-ety-wide terror that odors themselves caused disease. Urban planning shifted in the eighteenth and nineteenth centuries toward elaborate "deodorization" of cities: paving streets and replacing dirt paths with concrete to trap odors. In Manhattan it even prompted a grid-based street system that, it was thought, would encourage odors to race out of the city to the rivers, instead of settling into pleasant nooks and alleys. This surely reduces the dog's possible enjoyment of the smells inside the crevices of every fallen leaf and blade of grass paved over.

BRAMBISH AND BRUNKY

I used to be fooled by Pump's motionless posture when we sat outside together. One time, looking more closely at her, I saw that she was motionless but for one part: her nostrils. They were churning information through their caverns, ruminating on the sight before her nose. What was she seeing? The unknown dog who just turned the corner off the block? A barbecue down the hill, with perspiring volleyballers circling grilling meats? An approaching storm, with its fulminating bursts of air from distant climes? The hormones, the sweat, the meat—even the air currents preceding the arrival of a thunderstorm, upwardly moving drafts leaving invisible scent tracks in their wake—are all detectable, if not necessarily detected or understood, by the dog's nose. Whatever it was, she was far from the idle creature she'd seemed to be.

Knowing the importance of odor in a dog's world changed the way I thought about Pump's merry greeting of a visitor in my house by heading directly for his groin. The genitals, along with the mouth and the armpits, are truly good sources of information. To disallow this greeting is tantamount to blindfolding yourself when you open the door to a stranger. Since my guests may be less keen on the dog umwelt, though, I advise visitors to proffer a hand (undoubtedly fragrant), or kneel and let their head or trunk be sniffed instead.

Similarly, it is peculiarly human to chastise a dog for greeting a new dog in the neighborhood by smelling his rump. Our distaste for the notion of rump-smelling as a human social practice is irrelevant. For dogs, by all means, the closer the better. Dogs will communicate to each other if they are uninterested in being so intimately examined; interference may agitate one or both of them.

To understand the dog umwelt, then, we must think of objects, people, emotions—even times of day—as having distinctive odors. That we have so few words for smells restricts our imagination of the brambish, brunky diversity that exists. Perhaps, a dog can detect what a poet evokes: the "brilliant smell of water, / The brave smell of a stone, / The smell of dew and thunder . . ." (and definitely ". . . The old bones buried under . . ."). Probably, not all smells are good smells: as there is visual pollution, so is there olfactory pollution. Definitely, those who see smells must remember in smells, too: when we imagine dogs' dreaming and daydreaming, we should envisage dream images made of scents.

Since I've begun to appreciate Pump's smelly world I sometimes take her out just to sit and sniff. We have smell-walks, stopping at every landmark along our routes in which she shows an interest. She is *looking;* being outside is the most smelly, wonderful part of her day. I won't cut that short. I even look at photographs of her differently: where she once looked to be pensively staring in the distance, I now think what she's really doing is smelling some new exciting air from a far-flung source.

But I'm happiest of all to receive her greeting sniff of me, prompting her wag of recognition. I nuzzle into the scruff of her neck and sniff her right back.

Mute

Pump sits close to me and quietly pants, gazing at me: she wants something. On our walks she tells me when we've gone far enough and she is ready to go back: she hops up, pivots on her rear legs, then beelines back from where we came. I turn on the bathwater, turn to her with a smile, and her tail drops and wags low, her ears flattening on her head. All this talking and yet no talking at all.

There is a certain poignancy in describing animals as our "dumb friends"; in noting the "blank bewilderment" of a dog; in nodding at their "uncommunicating muteness." These are familiar ways of talking about dogs, who never respond in kind as we speak to them. No small amount of dogs' winsomeness is the empathy that we can attribute to them as they silently contemplate us. Still, these characterizations, while evocative, seem to me to be outright flawed in two ways. First, it is not the animals who desire to speak and cannot, I suspect; it is that we desire them to talk and cannot effect it. Second, most animals, and dogs in particular, are neither blank of expression nor in fact mute. Dogs, like wolves, communicate with their eyes, ears, tail, and very posture. Far from pleasantly silent, they squeal, growl, grunt, yelp, moan, whine, whimper, bark, yawn, and howl. And that's just in the first few weeks.

Dogs talk. They communicate; they declare; they express themselves. This comes as no surprise; what is surprising is how often they are communicating, and in how many ways. They talk to each other, they talk to you, and they talk to noises on the other side of closed doors or hidden in high grasses. This gregariousness is familiar to us: having a large roster of communications is consistent with being social, as humans are. Those canids such as foxes, who do not live in a social group, appear to have a much more limited range of things to say. Even the kinds of sounds foxes make are indicative of their more solitary nature: they make sounds that travel well over long distances. Dogs' staunch unmuteness is expressed through making sounds bellowed and whispered. Vocalizations, scent, stance, and facial expression each function to communicate to other dogs and, if we know how to listen, to us.

OUT LOUD

Two human beings stroll through a park chatting. They move with ease from commenting on the warmth of the air, to the nature of humans in positions of power, to expressions of mutual adoration, to reflections on past expressions of mutual adoration, to admonishment to observe the tree straight ahead. They do this primarily by making small, strange contortions of the shape of the cavities of their mouths, the placement of their tongues, by pushing air through the vocal tract and squeezing or widening their lips. Theirs is not the only communication going on. Over the course of a walk, the dogs by their sides may scold one another, confirm friendships, court each other, declare dominance, rebuff advances, claim ownership of a stick, or assert allegiance to their person. Dogs, like so many non-human animals, have evolved innumerable, non-language-driven methods to communicate with one another. Human facility at communication is unquestionable. We converse with an elaborate, symbol-driven language, quite unlike anything seen in other animals. But we sometimes forget that even non-language-using creatures might be talking up a storm.

What animals have are whole systems of behavior that get information from a sender (speaker) to a recipient (listener). That is all that is needed to call something a communication. It needn't be important, relevant, or even interesting information, but between animals it often is. Communication is only sometimes within our range of hearing, or even vocal: it is often made through body language—using limbs, head, eyes, tails, or the entire body—or even through such surprising forms as changing color, urinating and defecating, or making oneself larger or smaller.

We can spot a communication by noticing if, after one animal makes a noise or does an action, another responds to it by changing its behavior. Information has been imparted. What we'll miss, since we don't know the language of, say, spiders or sloths (though there are currently researchers trying to learn these communication systems), is those utterances that fall on deaf ears. Still, animals are constant gabbers. The discoveries of natural science over the last one hundred years have shown us the variety of guises in which this gabbing can appear. Birds twitter, peep, and sing songs—so do humpback whales. Bats emit high-frequency clicks; elephants, low-frequency rumbles. The wiggling dance of a honeybee communicates the direction, quality, and distance to food; a monkey's yawn conveys a threat. A firefly's flashes indicate his species; a poison-dart frog's coloration identifies his toxicity.

The kind we notice first is the one that most closely matches our own language: communication out loud.

DOG-EARED

Thunder outside. Pump's ears, velvet equilateral triangles that fold perfectly along the side of her head, prick into long isosceles. Head up, eyes to the window, she identifies the sound: a storm, a frightful thing. Her ears pivot back, flattened along her skull as if to hold them shut by their own force. I coo to her consolingly and watch her ears for feedback. The tips soften but she relaxes only slightly, still holding them tight against the roar.

Without prominent ears ourselves, we can envy dogs' proud ears. They come in a dazzling array of equally adorable variants: extremely long and lobular; small, soft, and perked; folding gracefully alongside the face. Dogs' ears may be mobile or rigid, triangular or rounded, floppy or upright. In most dogs, the *pinna*—the outer, visible part of the ear—rotates to better open a channel from the sound source to the inner ear. The practice of cropping ears, severing the pinnae to make floppy ears stand upright, long mandated in many breed standards, is becoming less popular. This designing of dogs, sometimes defended as reducing infections, has unknown consequences in auditory sensitivity.

By natural design, dogs' ears have evolved to hear certain kinds of sounds. Happily, that set of sounds overlaps with those we can hear and produce: if we utter it, it will at least hit the eardrum of a nearby dog. Our auditory range is from 20 hertz to

20 kilohertz: from the lowest pitch on the longest organ pipe to
an impossibly squeaky squeak.* We spend most of our time
straining to understand sounds between 100 hertz and 1 kilohertz, the range of any interesting speech going on in the vicinity. Dogs hear most of what we hear and then some. They can
detect sounds up to 45 kilohertz, much higher than the hair
cells of our ears bother to bend to. Hence the power of the dog
whistle, a seemingly magical device that makes no apparent
sound and yet perks the ears of dogs for blocks around. We call
this sound "ultrasonic," since it's beyond our ken, but it is within
the sonic range for many animals in our local environment.
Don't think for a moment that apart from the occasional dog
whistle, the world is quiet for dogs up at those high registers.
Even a typical room is pulsing with high frequencies, detectable
by dogs constantly. Think your bedroom is quiet when you rise
in the morning? The crystal resonator used in digital alarm
clocks emits a never-ending alarm of high-frequency pulses
audible to canine ears. Dogs can hear the navigational chirping
of rats behind your walls and the bodily vibrations of termites
within your walls. That compact fluorescent light you installed
to save energy? You may not hear the hum, but your dog probably can.

The range of pitches we are most intent on are those used in
speech. Dogs hear all sounds of speech, and are nearly as good as
we are at detecting a change of pitch—relevant, say, for understanding statements, which end in a low pitch, versus ques-

*In reality, few people hear equally well across this spectrum. With age, the
higher-frequency sounds, above 11–14 kilohertz, go undetected by the human ear.
This knowledge prompted the inspired design of a product with the teenager's
umwelt in mind. The device emits a 17 kilohertz tone—out of the range of most
adults' hearing, but unpleasantly audible to youngsters. Shop owners have used it as
a teenager repellent, to discourage loitering around their businesses.

tions, which in English end in a raised pitch: "Do you want to go for a walk(?)" With the question mark, this sentence is exciting to a dog with experience going on walks with humans. Without it, it is simply noise. Imagine the confusion generated by the recent growth of "up-talking," speech that ends every sentence with the sound of a question?

If dogs understand the stress and tones—the *prosody*—of speech, does this hint that they understand language? This is a natural but vexed question. Since language use is one of the most glaring differences between the human animal and all other animals, it has been proposed as the ultimate, incomparable criterion for intelligence. This raises serious hackles in some animal researchers (not thought of as a hackled species, ironically), who have set about trying to demonstrate what linguistic ability animals have. Even those researchers who may agree that language is necessary for intelligence have nonetheless added reams of results to the growing pile of evidence of linguistic ability in non-human animals. All parties agree, though, that there has been no discovery of a humanlike language—a corpus of infinitely combinable words that often carry many definitions, with rules for combining words into meaningful sentences—in animals.

This is not to say that animals might not understand some of our language use, even if they don't produce it themselves. There are, for instance, many examples of animals taking advantage of the communicative system of nearby unrelated animal species. Monkeys can make use of nearby birds' warning calls of a nearby predator to themselves take protective action. Even an animal who deceives another animal by mimicry—which some snakes, moths, and even flies can do—is in some way using another species's language.

The research with dogs suggests that they do understand

language—to a limited degree. On the one hand, to say that dogs understand *words* is a misnomer. Words exist in a language, which itself is product of a culture; dogs are participants in that culture on a very different level. Their framework for understanding the application of the word is entirely different. There is, no doubt, more to the words of their world than Gary Larson's *Far Side* comics suggest: eat, walk, and fetch. But he is on to something, insofar as these are organizing elements of their interaction with us: we circumscribe the dog's world to a small set of activities. Working dogs seem miraculously responsive and focused compared to city pets. It is not that they are innately more responsive or focused, but that their owners have added to their vocabularies types of things to do.

One component in understanding a word is the ability to discriminate it from other words. Given their sensitivity to the prosody of speech, dogs do not always excel at this. Try asking your dog on one morning to *go for a walk;* on the next, ask if your dog wants to *snow forty locks* in the same voice. If everything else remains the same, you'll probably get the same, affirmative reaction. The very first sounds of an utterance seem to be important to dog perception, though, so changing the swallowed consonants for articulated ones and the long vowels for short ones—*ma for a polk?*—might prompt the confusion merited by this gibberish. Of course humans read meaning into prosody, too. English does not give the prosody of speech syntactical leverage but it is still part of how we interpret "what has just been said."

If we were more sensitive to the *sound* of what we say to dogs, we might get better responses from them. High-pitched sounds mean something different than low sounds; rising sounds contrast with falling sounds. It is not accidental that we find ourselves cooing to an infant in silly, giddy tones (called *motherese*)—and might greet a wagging dog with similar baby talk. Infants can hear other speech sounds, but they are more interested in motherese. Dogs, too, respond with alacrity to baby talk—partially because it distinguishes speech that is directed *at* them from the rest of the continuous yammering above their heads. Moreover, they will come more easily to high-pitched and repeated call requests than to those at a lower pitch. What is the ecology behind this? High-pitched sounds are naturally interesting to dogs: they might indicate the excitement of a tussle or the shrieking of nearby injured prey. If a dog fails to respond to your reasonable suggestion that he come *right now,* resist the urge to lower and sharpen your tone. It indicates your frame of mind—and the punishment that might ensue for his prior uncooperativeness. Correspondingly, it is easier to get a dog to *sit* on command to a longer, descending tone rather than repeated, rising notes. Such a tone might be more likely to induce relaxation, or preparation for the next command from their talky human.

There is one celebrated dog whose word usage is exceptional. Rico, a border collie in Germany, can identify over two hundred toys by name. Given an enormous heap of all the toys and balls he has ever seen, he can reliably pull out and retrieve the one his owner requests. Now, putting aside why a dog might need two hundred toys, this ability is impressive. Children are hard-pressed to do the same task (and are only sometimes helpful in bringing things back). Even better, Rico can quickly learn a name for a new object, by process of elimination. Experimenters put a novel toy among familiar ones and asked him, using a word he had never heard before, to retrieve it. *Go get the snark, Rico.* One would be sympathetic if he looked bemused, and wandered back with a favorite toy in his chops. Instead, though, Rico reliably picked out the new toy: *naming* it.

Rico was not using language, of course, in the way we, or even young children, do. One can debate how much he was *understanding,* or if he was even doing anything other than showing a preference for the new object. On the other hand, he was showing an astute ability to satisfy the humans making various sounds by picking up the referents of those sounds. His ability might not indicate that all dogs are so able: Rico might be an unusually skilled word user*—and he is definitely unusually motivated by the praise received on retrieving the right toy. Still, even if he were the only dog who does this, it indicates that the dog's cognitive equipment is good enough to understand language in the right context.

It is not only the express content or sound of speech that carries meaning. Being a competent language user means understanding the pragmatics of usage: how the means, form, and context of what you say also affect the meaning of what you say. Paul Grice, a twentieth-century philosopher, famously described various "conversational maxims," known to us implicitly, that regulate language use. Their use marks you as a cooperative speaker; even their express violation is often meaningful. They include the charming maxim of relation (be relevant), the maxim of manner (be brief and clear), and maxims of quality (tell the truth) and quantity (say only as much as you need to).

On a good day, dogs mind all of Grice's maxims. Consider a dog who espies a roguish-looking fellow down the street. The dog may bark (relevant: the guy is roguish-looking) sharply (quite unambiguous), but only as long as the fellow is around (so the warning bark is currently true), and not more than a few

*Since the publication of Rico's successes, in 2004, other dogs (also border collies, for the most part) have been reported with vocabularies from eighty to over three hundred words: all names for various toys. You might have one of these prodigious vocabularians in your house.

times (relatively pithy). While dogs hardly qualify as competent language users, it is notably not because of their violation of the pragmatics of communication. It is only the smallness of their vocabulary and restricted use of words in combination that disqualifies them.

Many owners lament that, by contrast to Rico, their dogs are not terrific listeners—despite their broad range of audition. To be fair, canids do not rely on hearing as their primary sense. Relative to even our hearing, their ability to pinpoint where a sound is coming from is imprecise. They hear sounds unmoored from their origins. And just like us, they must bring attention to a noise to hear it best—first apparent in the familiar tilt of the head, to direct the ears slightly toward the sound source, or in radar-dish adjustments of the pinnae. Instead of being used to "see" the source of the sound, their auditory sense seems to serve an ancillary function: helping dogs find the general direction of a sound, at which point they can turn on a more acute sense, like olfaction or even vision, to investigate further.

Dogs themselves make a variety of sounds across a range of pitches or differing only by subtle alterations in tempo or frequency. They are downright noisy.

THE OPPOSITE OF MUTE

Her slow, light panting, mouth open partway, tongue purple and wet and perfect. Pump's panting was a conversation in and of itself—I always felt talked-to when she panted at me.

The cacophony of a packed dog run seems at first pass to be an undifferentiated racket. With closer attention, though, one can distinguish shouts from cries; yelps from barks; and play barks from threatening barks. Dogs make sounds both intentionally and inadvertently. Both kinds may hold information, the minimum requirement to call an aural disturbance a "communication" rather than simply "noise." What is interesting for scientists is determining the meaning of that information. Given the way dogs wield these noises, there is no doubt that they have different meanings.

Countless hours of researchers' lives spent listening to animals shout, coo, click, groan, and scream has led to the discovery of some universal features of sound signals. They either express something about the world—a discovery, a danger—or something about the signalers themselves—their identity, sexual status, rank, membership in a group, fear, or pleasure. They effect a change in others: they may decrease social distance between the signaler and those around him, calling someone closer; or increase social distance, frightening someone away. In addition, sounds may serve to cohere a group (in defense from a predator or intruder, for instance) or they may elicit maternal or sexual affiliation. Ultimately, all these purposes for making sounds make evolutionary sense: they aid the animal in securing its survival or the survival of its relatives.

What, then, are dogs saying, and how are they saying it? The *what* is answered by looking at the context of making a sound. The context includes not just the sounds around it but also the means: a screamed word winds up meaning something different than one intoned with a sultry whisper. A sound a dog makes while wagging merrily means something different than the same sound delivered through bared teeth.

The meaning of an uttered sound can also be identified by looking at what those who hear it do. Although human responses to an utterance (say, *How are you?*) may range from the appropriate (*I'm well, thanks*) to the seeming non sequitur

(*Yes, we have no bananas*), there is reason to believe that dogs, and all non-human animals, respond ingenuously. In many cases, a sound will have a reliable effect on those in the vicinity: think *Fire!* or *Free money!*

The *how* of sound signaling is simple with dogs. Most of the sounds dogs make are oral: using or coming out of the mouth. At least, these are the sounds that we know about. These vocal sounds might be voiced, with vibration in the larynx—the airway used for breathing—or may be expiratory—part of an exhalation. Others are entirely unvoiced but use the mouth, such as the mechanical sound of tooth-snapping. Vocal sounds vary from one another along four easily audible dimensions. They vary in pitch (frequency): whines are nearly always high-pitched, while growls are low-pitched. Try and squeal out a growl and it becomes something else. They vary in duration: some are uttered once, quickly, lasting less than half a second; others are protracted sounds or are repeated again and again. Sounds vary in their shape: some are pure tones while others are more fractured, fluctuating or rising and falling. A howl has little variation for long periods, while barks are noisy, changeable sounds. Finally, they vary in loudness or intensity. Moans don't come in loud and yelps don't come in a whisper.

WHIMPERS, GROWLS, SQUEAKS, AND CHUCKLES

She sees I'm almost ready. With her head fixed on the ground between her paws, Pump follows me with her eyes as I cross the room gathering my bag, a book, my keys. I scratch her around her ears in consolation and break for the door. She lifts her head and makes a sound: a plaintive yelp. I freeze. A look back and she hurries over, wagging. Okay, then; I guess she'll come with me.

The paradigmatic dog sound is the bark, but barks do not form the preponderance of most dogs' daily noisemaking, which includes high and low sounds, incidental sounds, even howls and chuckles. High-frequency sounds—cries, squeals, whines, whimpers, yelps, and screams—occur when the dog is in sudden pain or needs attention. These are some of the first sounds a puppy will make, which clues us in to their meaning: they tend to attract attention of the mother. A *yelp* might come out of a puppy who was just stepped on, or who has wandered off. Deaf and blind, it is easier for mom to find her pup than vice versa. Having been reunited, some continue to yelp, winding down off their crying jag, when carried by their mothers. Yelps are different than *screams,* which in wolves prompt the mother to groom the pup, providing the contact that is necessary for normal development. *Cries* and *squeals* may be ignored by the mother, and so a particular squeal may be a less specifically meaningful utterance and instead a general-purpose sound used simply to see how others respond.

Low *moans* or *grunts* are also very common in puppies, and seem not to be signs of pain but rather a kind of dog purr. There are snuffling moans and sighing moans—what some call "contentment grunts," and they all seem to mean about the same thing. Pups moan when they are in close contact with littermates, their mother, or a well-known human caretaker. The sound might be simply a result of heavy, slow breathing, which indicates it might not be intentionally produced: there is no evidence that dogs moan on purpose (neither is there evidence that they do not; neither has been proven). But whether they do or not, moans probably function to affirm the bond between family members, whether heard as a low vibration or felt through skin-to-skin contact.

The rumble of a *growl* and the steady ominous *snarl,* you won't need to be told, are aggressive sounds. Puppies do not tend to produce them, as puppies do not tend to initiate aggression. Part of what makes them aggressive is their low pitch: they are the kind of sounds that would come out of a large animal, rather than the high-pitched squeals of a small one. In an antagonistic (what in biology is called *agonistic*) encounter with another animal, a dog wants to appear to be the bigger, more powerful creature—so he makes a big-dog sound. By making higher-pitched sounds, an animal sounds, simply, smaller: a friendly or appeasing noise, by contrast. Though aggressive in intent, growls are still *social* sounds, not just utterances produced when a dog feels fear or anger: for the most part, dogs do not growl at inanimate objects,* or even at animate objects that aren't faced or directed toward them. They are also subtler than we think: distinct growls, from rumble to nearly roaring, are used in different contexts. The growl of tug-of-war may sound fearsome, but it is nothing like the possessive warning snarled over a treasured bone. Play these growls back over a speaker set up right in front of a desirable bone and dogs in the vicinity will avoid the bone—even with no dog in sight. But if the speaker growls only play or stranger growls, nearby dogs go ahead and grab the unguarded bone.

---

*Except when it is animated: a discarded plastic bag being tumbled down a city sidewalk by a breeze can provoke growls, caution, and occasional attacks by alarmed dogs. Dogs can be animistic, just as humans are in infancy: trying to make sense of the world by attributing a familiar quality (of life) to unknown objects. My plastic-bag-growling dog is in good company: Darwin described his own dog treating an open parasol moving in the breeze as a living thing, barking at and stalking it. And Jane Goodall has observed chimpanzees making threatening gestures toward thunderclouds. I've been known to fulminate thundercloud-ward myself.

Incidental sounds of dogs are sometimes produced so reliably in certain contexts that they have become effectively communicative. The *play slap,* an audible landing on the two forefeet at once, is an inevitable part of play. It conveys sufficient exuberance that it can be used by itself to ask a dog to play with you. Some dogs *chatter* their teeth in anxious excitement, and the clicking of teeth serves as a warning that the dog is wary. An exaggerated *shriek* on being nosed or bitten roughly in play can even become a ritualized deception, a way to get out of a social interaction that is making the dog uncertain. The *snuffling* sound created when reaching the head vertically up and sniffing for food around a human mouth can become not just a search for food but also a request for food. Even the noisy breathing created by lying so close as to have the nose pressed against another body comes to indicate a state of contented relaxation.

If you live with a hound, you are familiar with the *howl.* From a staccato baying to a mournful wail, howling in dogs seems to be a behavior left over from their ancestors, living in social packs. Wolves howl when separated from the group, and also when setting out with the group for a hunt or in reunion afterward. A howl when alone is a communication seeking company; howling together may be simply a rallying cry or celebration of the group. It has a contagious component, leading others in the vicinity to pick it up in an impromptu fugue. We do not know what they are saying, to each other or to the moon.

The most social of human sounds is the cackling laugh rumbling across the room. Do dogs laugh? Well, only when something is terrific fun. Yes, dogs have what has been called a laugh. It is not identical to human laughter, the spontaneous sounds spit out in response to something funny, surprising, or even frightening. Nor is it as variable as the cackles, giggles, and twitters that we produce. The dog laugh is a breathy exhalation that sounds like an excited burst of panting. We could call it *social panting:* it is a pant only heard when dogs are playing or trying to get someone to play with them. Dogs don't seem to laugh to themselves, off sitting in the corner of the room, recollecting how that tawny dog in the park outsmarted her human this morning. Instead, dogs laugh when interacting socially. If you have played with your dog, you have probably heard it. In fact, doing your own social panting toward a dog is one of the most effective ways to elicit play.

Just as our laughs are often inadvertent, reflexive responses, so may dog laughs be: simply the kind of panting that results when you're throwing your body around in play. Though it might not be under the control of the dog, social panting does seem to be a sign of enjoyment. And it may induce enjoyment— or at least alleviate stress—in others: playing a recording of the sounds of dog laughter at animal shelters has been found to reduce barking, pacing, and other signs of stress in the dogs housed there. Whether mirth feels like what it does in humans is yet to be studied.

WOOF

I can remember the first time a bark came out of Pump, when she was maybe three years old. She'd been so quiet until then, and then one day, after spending time with her barky German shepherd friend, a bark popped out of her. It was bark*like* more than a bark, as though a sound that stood for a bark but wasn't itself the real thing: a well-articulated *rurph!* accompanied by a little leap off her front legs and a madly wagging tail. She refined this splendid display somewhat through the years, but it always felt like a new dog thing she was trying on.

It is regrettable that barks tend to be such loud affairs. The bark is shouted. While a calm conversation between the two strollers in the park might register about 60 decibels, dog barks begin at 70 decibels and a stream of barks may be punctuated with spikes to 130 decibels. Increases in decibels, the unit of measurement of the loudness of sounds, are exponential: an increase of 10 decibels describes a hundredfold rise in the experience of the strength of a sound. One hundred and thirty decibels is up there with thunderclaps and plane takeoffs. The bark is momentary, but it is a moment of displeasure for our ears. The reason this is regrettable is that there is, most dog researchers agree, much information in those barks. Given the relative scarcity of barking in wolves, some theorize that dogs have developed a more elaborate barking language precisely in order to communicate with humans. If we consider barks as all cut from the same cloth, though, they are likelier to annoy than to communicate.

Researchers might not call barks "annoying," but they call them "chaotic" and "noisy." "Chaotic" is a good description for the variability in the kinds of sounds within each bark; "noisy" means not just *disagreeably loud* but also *having fluctuations in its structure.* Barks are loud, and different barks have varying numbers of harmonic components, depending on the context in which the bark is used.

Still, of the sounds dogs make, barks come closest to speech sounds. The dog's bark is, like the phonemes of speech, produced by vibrations in the vocal folds and air flowing along the folds and through the mouth cavity. Perhaps because they are in overlapping frequencies—from 10 hertz to 2 kilohertz—with the sounds of speech, we are inclined to look for speechlike meaning in them. We even name the bark using phonemes from our language: the dog "woofs," "rufs," "arfs," or (although no dog I know says it) "bow-wows." The French hear the dog "ouah-ouah"; Norwegian dogs *voff-voff;* Italian dogs go *bau-bau.*

Some ethologists think that barking is not fundamentally communicating anything, though: that it is "ambiguous" and "meaningless." This view is encouraged by the difficulty of deciphering what the meaning of the barks could be, since sometimes dogs bark without an obvious prompt or audience, or continue to bark long after any message therein would have been conveyed. Think of the dog barking continuously, dozens of times in a row, in front of another dog: If there is meaning in that bark, would not one or two repetitions do to convey it?

This strikes at the heart of the trouble in determining the subjective experience of an animal of which you cannot ask questions. Each moment of an animal's behavior is scrutinized for its meaning. Surely few human actions could bear such scrutiny and yield a correct assessment of the human. If you were to videotape me practicing, at home in front of my dog, a speech I am to deliver later in the day, you might well conclude that (a) I believe the dog can understand what I'm saying, or (b) I am talking to myself. In either case, (c) holds: the noises I am making appear not to be classically communicative—for I do not have an audience who can understand me. Similarly, examples of poor communication by a dog may seem to undermine the notion that dogs communicate at all. But most researchers think that barks do have meaning, albeit one dependent on the context and even on the individual. Barking, especially alarm barking, is one of the clearest distinctions between dogs and other canid species. Wolves bark to convey alarm, but rarely, and they make more of a "woof" sound than anything like the protracted dog barking with which we are familiar. Dogs do not just bark more than wolves; they have developed numerous variations on the theme.

There are a handful of distinguishable barks, used reliably in a handful of distinguishable cases. Dogs bark to get attention, to warn of danger, in fear, as a greeting, in play, or even out of loneliness, anxiety, confusion, distress, or discomfort. The meaning is in the context of their use, but not only in the context: spectrograms of dog barks show that they are mixtures of the tones used in growls, in whimpers, and in yelps. By altering the prevalence of one tone over the others, the bark takes on a different character—a different gist.

Early research into dogs' vocalizations concluded that all dog barking was attention-getting barking. In fact they do attract attention, assuming someone is close enough to hear them. But recent studies have made more subtle discriminations between barks. While in some way all barks come down to some manner of "attention-getting," one might as well say that we speak in order to be heard: true, but incomplete. For instance, when experimenters analyzed the spectrograms of thousands of dog barks during one of three contexts—a stranger ringing the doorbell, being locked outside, or in play—they found three distinct types of barks.

Stranger barks were the lowest in pitch and the harshest: they are nearly spat out. Less variable than the other types, stranger barks are well designed to send a message over a distance, something necessary if caught in a threatening situation alone. They can also be combined into "superbarks," concatenations of barks that together last much longer than the duration of barking in other contexts. The end result is a bark that most human listeners find to be aggressive.

The *isolation barks* tended to be higher-frequency and more variable: some ranged from loud to soft and back again, some went from high to low. These barks are lobbed into the air one by one, sometimes with great intervals between them. They sound "fearful," people tend to say.

Play barks, too, are high-frequency, but they happen more often one after the other than the isolation barks. They're directed at someone else, unlike isolation barks: at a dog or human playmate. There is considerable individual variation, of course: not every dog barks alike. The stranger bark of a small dog may come out as *rar, rar* or *raoaw, raoaw,* while a larger dog emits a capital-r *Rumph.*

These differences between bark types make evolutionary sense: the lowest sounds are used in threatening situations (again to appear bigger); higher sounds are entreaties—to friends, for companionship—and as such are submissive requests, not warnings. Differences between individual barkers indicate that barks might be used to affirm a dog's identity, or reveal an association with a group (even the group *me and the woman at the end of my leash,* rather than *these dogs I'm frolicking among*). And barking together with others may be a form of social cohesion. Barking can be contagious, like the howl: one dog barking might prompt a chorus of barking dogs, all joined in their shared noisiness.

BODY AND TAIL

When we approached people on the street, Pump set all her senses to looking; if she recognized them, her head would lower ever so slightly—looking up coyly as though over reading glasses—and she would wag her tail low. This was quite different from her approach of a dog she was smitten with: all upright, tail high, posture impeccable, wags soldierlike in their rhythm— or a dog friend—a looser, janglier approach, and even an open-mouthed grab toward their face, or a gentle bump with her hip along their body.

You may be sitting down right now, folded into a comfy chair—or perhaps you're standing, straphanging on a train to work, book scrunched against another commuter's back. Most likely you don't *mean* anything by your sitting or standing, or when you walk or lie supine: it's just a posture of convenience or comfort. But in other contexts our very posture conveys information. A catcher crouches: he's prepared for a pitch. A parent crouches and opens his arms: he's inviting a child for a hug. Running when someone you know approaches, you suddenly stop and greet them; standing still when someone you know approaches, you suddenly turn and run. There can be meaning simply in the vigor or slouch of your body. For an animal with a limited vocal repertoire, posture is ever more important. And it appears that dogs use specific postures to make very specific statements.

There is a language of the body, formed of phonemes made from rumps, heads, ears, legs, and tails. Dogs know how to translate this language intuitively; I learned it after watching hundreds of hours of dogs interacting with each other. We must look like such stiffs to dogs, who can express everything from playfulness to aggression to amorous intent by changing the shape of their body and its altitude. By contrast, we are inhibited straight-backeds, mostly stationary or traveling forward with little excess movement. Occasionally—heavens—we turn a head or arm flamboyantly to the side.

> But man himself cannot express love and humility by external signs, so plainly as does a dog, when with drooping ears, hanging lips, flexuous body, and wagging tail, he meets his beloved master. —Charles Darwin

For dogs, posture can announce aggressive intent or shrinking modesty. To simply stand erect, at full height, with head and ears up, is to announce readiness to engage, and perhaps to be the prime mover in the engagement. Even the hair between the shoulders or at the rump—the *hackles*—may be standing at attention, serving not just as a visual signal of arousal but also releasing the odor of the skin glands at the base of the hairs. To exaggerate the whole effect, a dog might stand not just up but *over* another dog, head or paws on his back. That's about as declarative a statement as you can make that you are feeling dominant. The opposite body posture, crouching with head down, ears down, and tail tucked away, is submissive. To lie all the way down and expose the belly is even more so.*

This principle of *antithesis*—that opposing postures communicate opposing emotions—describes much of the expressive scope of dogs. Facial expressions, most visible in the mouth and ears, mind this principle, too. The mouth sweeps from closed to open and relaxed, to open with lips raised, snout wrinkled, and teeth bared. A dog's "grin," with jaw closed, is submissive; as the mouth is opened, the arousal increases; and if the teeth are exposed, the look gets aggressive. Coming full circle, a wide-open mouth with teeth mostly covered—a yawn—is not a sign of boredom, as often assumed by analogy with our own yawning; instead, it may indicate anxiety, timidity, or stress, and is used by dogs to calm themselves or others. The ears can also go through these gymnastics: they can be pricked, relaxed and down, or folded tightly along the head. Eyeballing another dog directly can be threatening or aggressive; by contrast, looking away is submissive—an attempt to quell one's own anxiety and

---

*Surprisingly, dogs mind each other's posture more than their height: dogs do not read simply being taller as being dominant or confident. As we'll see later, it's not quite right to say, as is often said of a bravely forward small dog, *he thinks he's a big dog:* Actually, he does not—he knows it is posture that matters.

the other dog's arousal. In other words, in each case there is a range from one extreme to another, representing variation in intensity along an emotional continuum, from relaxed to aroused in fear or alarm.

None of these is a static symbol—or, if it is, its being static is meaningful. Holding an erect posture, motionless, is a quiet way of putting an exclamation point on the posture. It exaggerates the tenseness of the communication. For the most part, postures are taken, and moved through. The tail, especially, is a limb of movement. It is to science's great discredit that no one has done a thorough investigation of the meaning of every wag of the dog's tail.

As a puppy her tail was trim, an arrow of soft black fur. This turned out not to be the destiny of her tail at all: it grew into an incredible banner of a tail, with extravagant feathering that matted and gathered leaves. It was bent at the tip from a disagreement with a car door when she was young. She brandished it when excited or delighted, curved to a sickle with the tip pointing at her back. When lying down, she drummed it happily on the ground at my approach. Her tail registered her exhaustion in a low-hanging straight pose; her disinterest in a nosy dog by tucking between her legs. Most of the time as we walked together, it hung loosely down, curving jauntily toward its tip, and merrily swished to and fro. I loved to approach her slowly, stalking her, and prompt her tail to quiver into wagging.

One of the difficulties in deciphering the language of tails is the great variation in tails among dogs. The flamboyant plumage of a golden retriever contrasts mightily with the tight corkscrew of the pug. Dogs wear tails long and rigid, stumpy and curled, hanging heavily or perpetually perked. The wolf tail is in some ways an average of the various breed tails: it is a long, slightly feathered tail, held naturally slightly down. Early ethologists who did a reckoning of wolf tail postures identified at least thirteen different tail carriages, conveying thirteen distinguishable messages. As per the antithesis thesis, tails held high indicate confidence, self-assertion, or excitement from interest or aggression, while low-hanging tails indicate depression, stress, or anxiety. An erect tail also exposes the anal region, allowing a bold dog to air his odor signature. By contrast, a long tail held so low as to curl back between the legs, closing off the rump, is actively submissive and fearful. When a dog is simply waiting around, his tail is relaxed, hanging low, dropped down but not rigid. A tail gently lifted is a sign of mild interest or alertness.

But it is not as simple as tail height, for the tail is not just held, it is wagged. Wagging does not translate as simple happiness. A high, stiffly wagging tail can be a threat, especially when accompanied by an erect posture. Quickly wagging a dropped, low tail is another sign of submission. This is the tail of the dog who has just been caught finishing off the last of your shoes. The vigor of the wag is roughly indicative of the intensity of the emotions. A neutral tail wagging lightly is interested but tentative. A loose, lively whisking tail accompanies the nose-led search for a ball lost in high grasses or an odor trail discovered on the ground. The familiar happy wag is incredibly different from all of these: the tail is held above or out from the body and strongly draws rough arcs in the air behind it. Unmistakable delight. Even non-wagging is meaningful: dogs tend to still their tails when attending carefully to a ball in your hand or waiting for you to tell them what's happening next.

Researchers interested in the brain of the dog accidentally discovered something about the tail of the dog along the way: the dog wags asymmetrically. On average, dogs' wags tend more strongly to the right when they suddenly see their owners—or even anything else of some interest: another person, or a cat. When presented with an unfamiliar dog, dogs still wag—more that tentative wag than the happy wag—but tending to the left. You might not be able to see this in your own dog unless you watch them wag in slow-motion video playback (which I highly recommend)—or unless your dog is one of those who wag less back-and-forth than round-and-round, inclining to the side. Consider yourself lucky to be wagged at with such clear-cut enthusiasm.

> Pump does a full-body shake: it starts in her head and rolls down her body, shimmering out through her tail. It is like a punctuation mark that has yet to be discovered. She shakes to end an episode, when she's unsure, and sometimes when she's just ambling along.

The dog uses his body expressively: communication writ through movement. Even the moments between interactions are marked by movement: as when a dog does a full-body shake, his skin twisting over his frame, to indicate his finishing one activity and moving on to another. Not all dogs have hackles that visibly raise with pique, long tails to ostentatiously wag, or ears that raise with interest. The fabulously ropy-furred komondor approaches other dogs with what we must assume is his head, but neither eyes nor ears are visible underneath his long locks. In breeding dogs to have particular looks that we find agreeable, we are limiting their possibilities for communicating. Just as we might expect, but would rather not confront, a dog with a docked tail has, thereby, a docked repertoire of things he can say.

Research looking at the range and rate of signals used by ten physically dissimilar breeds found just this. Comparing the behavior of dogs from the Cavalier King Charles spaniel to the French bulldog to the Siberian husky, there was a clear relationship between the breed appearance and the number of signals they used. Those animals that had been most changed physically in domestication from wolves—the King Charles, at the extreme—sent the fewest signals. These *pedomorphic* or *neotenous* dogs, who retain more features of juvenile members of canid species into adulthood, look least like adult wolves. The huskies, which have the most wolflike features and are genetically closer to *Canis lupus,* do the most wolflike signals.

Given that many bodily signals provide information about one's status, strength, or intent, the necessity for dogs to send these signals is presumably diminished in a world where humans chaperone dogs through life. But the same signals used to convince a dominant animal of benign intent may also be used to communicate information to humans. Walking through the city, I turn a blind corner and nearly step on an unfamiliar dog pulling on a long leash. Seeing me, she crouches, wags her tail furiously between her legs, and licks toward my face. It may have begun as a submissive gesture, but now it is adorable.

INADVERTENT AND INTENT

After sleeping late and enduring the slow pace of my morning rituals, Pump's first move when we get outside never varies. She takes two steps out the door and unceremoniously squats. She crouches deeply, fully committed to the pose, with only her tail—curled high out of the way—pulling her body up. The torrent of urine released (surely record-breaking this time) seems accompanied by a relaxing of the muscles of her face—and by a growing guilt on my own that I made her wait so long. She watches the stream of her own urine wandering by her as it finds cracks in the sidewalk in which to divert itself cleanly.

As much as is being said with a bark, snarl, or wagging tail, vocalization and posture are not the only media of communication for dogs. Neither is a match for the informational possibilities of smells. Urination, as we saw earlier, is the means of odorant communication most conspicuous to us. It might be hard to believe that the release of the bladder is a "communicative act" right up there with a polite conversation between friends or a politician orating before his constituents. At some level, it is like both of these: it is part of normal dog sociality, and it can also be a bellowing self-promotion writ on a hydrant.

You might balk at calling the moist message left high on a lowly fire hydrant the same kind of communication that humans use—and not just because they are talking out of their rumps, not their faces. Crucially, we communicate (most of the time) with *intention:* rather than yammering out loud to our own left hand, we tend to direct our communications to other people— people who are near enough to hear us, not otherwise distracted, who know the language, and can understand what we're saying. Intention distinguishes communication done with others in mind from the automatic *oof!* uttered on being hit in the belly, blushing at a compliment, a mosquito's constant buzzing, or the unmindful bit of information imparting done by traffic lights and flags at half-mast.

Urine marking is peeing with intention. The morning's blissful release relieves the strain on the bladder, but most of the time some is held back for later use in marking. Presumably the urine is the same urine: there's no evidence of an independent channel or means by which to modify the odor they are emitting. But marking behavior differs in a few key ways. First, in most adult males, and some gender-bending females, marking is characterized by a prominently raised leg. There are individual and contextual variations on the so-called "raised-leg display," from a modest retraction of the rear leg up toward the body, to hoisting the leg up above the hip points, above vertical—certainly also a visual display for any other dogs in the vicinity. Both allow for a directional flow of urine, aimed to land at a conspicuous site. (One can squat and mark, too, although it is a quieter affair, perhaps for messages better whispered than shouted.)

Second, the bladder is not emptied when marking; urine is doled out a little at a time, allowing for greater distribution of scent over the course of the dog's travels. If you've left your dog indoors long enough that he races outside to squat, this urgency might preempt his ability to cache some urine for later marking. Thus the fruitless raised-leg displays you may witness, waving dryly at bushes, lampposts, and trash cans.

Finally, dogs usually urine mark only after spending some time sniffing the area. This is what elevates the odor exchange from Lorenz's notion of flag planting to a kind of conversation. Researchers keeping careful tally of dogs' marking behavior over time found that who has marked before them, the time of year, and who is nearby all affect where and when they mark.

Interestingly, these message bouquets aren't left indiscriminately: not every surface is marked. Watch a dog sniff his way down the street: he will sniff more locations than he will squirt. This indicates that not every message is the same—and the message this dog will leave may be intended for certain audiences only. Countermarking—covering old urine with new—is a common behavior of male dogs, when the old urine is that of less dominant male dogs. Everyone's marking increases when there is a new dog around.

If it is not territorial, what is the message in the mark? The first hint is that puppies don't urine mark: the communication must have to do with adult concerns. From the position of the anal glands and the compounds in the urine, we know that they are at least saying something about who they are: their odor is their identity. This is a fine message, but it is probably fairly unintentional. I may communicate something about who I am by merely walking into a room and being seen, but the very fact of my person is not a continuous, intentional communication about my identity (except when I was a kid and dressed to be seen).

What does look intentional in this communication is that dogs don't bother to say it if there's no one else around. Dogs who are kept penned by themselves spend very little time marking. The males rarely lift a leg to urinate, and neither sex bothers to deposit just a small amount. Dogs kept in similar-sized enclosures with other dogs mark much more frequently, and they mark regularly, every day. The Indian feral dogs marked to audiences—and audiences of the opposite sex. This makes sense if the message conveyed is about sex: seeking it oneself, or declaring oneself fit to be seeked. They did the most raised-leg displays (even without urinating) when other dogs were present. A leg held high only gets someone else's attention if someone else is already there to attend to it.

It also makes sense if the mark is communication for communication's sake: a comment, an opinion, a strongly held belief. There is no scientific evidence that it is so, but it is consistent with communication done only to an audience. Researchers have found that dogs raised in isolation make many fewer communicative noises than those raised with other dogs. When finally around others, though, they begin producing vocalizations at the same rate that the socialized dogs do. In other words, they speak when there is someone to speak *to*.

Just as they mark with intention, so too do dogs read intention in our markings: in our gestures. As we will see in the next chapters, they interpret the body language of humans with the attention they bring to reading each other. As a young child toddles toward a treasured toy, a dog can see where she is going and get there first. A turn of the head in thought garners little attention, but a turn of the head that looks at the door—there is intention in that turn. And dogs know it. They realize that there is a difference between gazing toward the door and turning to look at the clock on the wall; they can distinguish a finger pointing toward hidden food, and a point done while lifting one's arm to check a watch. We speak loudly with our bodies.

A confession: a dog has dictated this entire chapter to me. She sat by my chair, head on my foot, and patiently waited while I struggled to translate her words to the page. It is from her that the insights of the book come, from her that evocations spring, from her that the scenes and images and umwelt emerge.

Alas, it is not quite so. But to see the remarkable number of volumes purportedly written by dogs one must imagine that this is what we all want: the story straight out of the dog's mouth—but in our native tongue, of course. At the end of the nineteenth century, a peculiar kind of autobiography began to appear in bookshops: it was the "memoir" of your cat, your old dog, or the animal gone missing in that winter storm. This form, narrated by talking animals, could be considered the first prose attempt to get at the point of view of the dog. When I read one of these—and there are many to choose from, among even such writers as Rudyard Kipling and Virginia Woolf—a strange discontent washes over me. It's a sham: there is no perspective of the dog in them. Instead, it is a dog with the human's voice box transplanted to the dog's snout. Imagining that dogs' thoughts are but cruder forms of human discourse does the dog a disservice. And despite their marvelous range and extent of communication, it is the very fact that they do not use language that makes me especially treasure dogs. Their silence can be one of their most endearing traits. Not muteness: absence of linguistic noise. There is no awkwardness in a shared silent moment with a dog: a gaze from the dog on the other side of the room; lying sleepily alongside each other. It is when language stops that we connect most fully.

Dog-eyed

It takes all of six seconds for Pump to go from masterful to maladroit. In the first five she flawlessly navigates the brambles and bushes and thick-trunked trees that web the opening of the forest into field to catch a fast-moving tennis ball. It thonks off a tree and she's there to nearly vacuum it into her mouth. An apparition of a dog tears in out of nowhere, a racing blur of white fur and bark. Pump notes him and hurtles away, evading this stealer of tennis balls. In that sixth second, she stops, suddenly adrift. She's lost track of me. I watch her search: erect body, head high. I'm within sight; I smile at her. She looks at and past me, not seeing me. Instead, she spots the large, limping, heavy-coated man who came through with the white tear. She takes off after him. I must run to retrieve her. The moment before Pump was all-seeing; now she's a fool.

There is an intrinsic ranking of the modes by which we humans sense the world—and vision is winner by a long shot. Eyes arouse great interest in human psychologists; they betoken much more than one might imagine just from their physical form. However pretty a nose one might have, however close the forehead is to the brain, neither our noses nor foreheads nor cheeks nor ears are granted such importance.

We are visual animals. There's barely a challenge for second, either: audition is part of nearly every experience we have. Olfaction and touch might duke it out for third, and taste runs a distant fifth. Not that each of these isn't important to us on any particular occasion. The loveliness of presentation of, say, a tiered wedding cake would be undercut if vinegar replaced the anticipated taste of unmitigated sweetness. Or if any odor besides that of baked goods emanated from the cake—or if the first bite was not soft and yielding, but crunchy or slimy. Still, on most occasions we first direct our gaze to a new scene or object. If we notice something unusual or unexpected on the sleeve of our jacket, we turn to examine it with our eyes. Vision would have to really fail to provide any information before we decide to learn about it by inhaling it closely or taking a bold lick.

The order of operations is turned upside-down for dogs. Snout beats eyes and mouth beats ears. Given the olfactory acuity of dogs, it makes sense that vision plays an accessory role. When a dog turns his head toward you, it is not so much to look at you with his eyes; rather, it is to get his nose to look at you. The eyes just come along for the ride. You may be a recipient of an imploring gaze by a dog across the room right now. But can dogs even see what we do?

In many ways, the visual system of a dog—a subsidiary means to look at the world—is very much like our own. Its demotion behind other senses, in fact, may allow dogs to see details with their eyes that we overlook with ours.

One might well ask what a dog would even need eyes for. They can navigate and find food with their remarkable noses. Anything that needs closer examination goes right in the mouth. And they can identify each other through that sensory apparatus squished between their mouth and nose, the vomeronasal organ. As it turns out, they have at least two critical uses of their eyes: to complement their other senses and to see us. The natural history of the dog eye, seen in the story of their forebears, wolves, explains the context in which their vision evolved. It is a happy and transformative side effect that this has made them good watchers of human beings.

Just one element of the lives of wolves goes a long way to explaining the eyes they have evolved: eating. Most of their food runs away. Not only that, it is often camouflaged or living in the relative safety of herds. It is active—and thus findable— at dusk, dawn, or night. So wolves, like all predators, have evolved in response to their prey. As important as scent is, it cannot serve as the only indicator of the presence of prey, as air currents send odors on circuitous paths before they reach the nose. Odors are volatile: if smell lies on a surface, a sensitive nose can track it specifically; but if it is on the wind, the odor appears more like a cloud that could have come from one of a thousand sources. Rapidly moving prey outrun their own odor. Light waves, by contrast, are transmitted reliably through open air. So after catching a whiff, wolves use their sight to locate their prey. Many prey animals are camouflaged to blend in with their environment. This camouflage is betrayed in motion, however. So wolves are adept at spotting a change in the visual scene that indicates that something is moving. Finally, prey animals are often active at dusk or dawn, a compromise of lighting: easier to hide, harder to see. In response, wolves developed eyes that are especially sensitive in low light, and are especially good at noticing motion in that light.

Her eyes are deep pools of brown and black. It is hard to see which way they gaze, they are so dark—but it also makes any glimpse of her irises delightful, as though seeing inside her soul. Her eyelashes only became apparent when they grayed. Her eyebrows are also essentially invisible, but the effect of their moving—as with her head on the floor, to follow me walking across the room—is visible. In sleep, in dreams, her eyes scanned the world under their lids. Even closed, the lids reveal a bit of pink peeking out, as though she were keeping prepared to open her eyes at once should something important happen nearby.

At first glance these prey-tracking eyes look much like ours: viscous spheres fitted in sockets. Our eyes are about the same size as dogs'. Despite the fact that dogs' heads vary so significantly in size (four Chihuahua heads would fit in a wolfhound's mouth—not that anyone would deign suggest verifying such a thing), eye size barely varies between breeds. Small dogs, like puppies and infants, have large eyes relative to the size of their heads.

But small differences between the eyes of humans and dogs immediately become apparent. First, our eyes are smack in the front of our face. We look forward, and images in the periphery fall away to darkness around our ears. While there is variation among dogs, most dogs' eyes are situated more laterally on their heads in the manner of other quadrupeds, allowing a panoramic view of the environment: 250–270 degrees, as contrasted to humans' 180 degrees.

If we look a little closer, we discover another key difference. The superficial anatomy of our eyes gives us away: it shows where we're looking, how we're feeling, our level of attention. While dog and human eyes are similarly sized, our pupil size—the black center of the eye that lets light in—varies considerably when we are in a darkened room or are aroused or fearful (expanding to up to 9 millimeters wide) or in the bright sun or are highly relaxed (contracting to 1 millimeter). Dogs' pupils, by contrast, are relatively fixed at about 3 to 4 millimeters, regardless of the light or the dog's level of excitement. Our irises, the muscles that control pupil size, tend to be colored to contrast with the pupil, blue or brown or green. Not so in most dogs, whose eyes are often so monochromatically dark that they remind me of bottomless lakes, repository of all attributions of purity or desolation we might ascribe to them. And the human iris sits amidst the sclera—the white—of the eye, while many (but not all) dogs have very little sclera. The anatomical sum effect is that we can always see where another person is looking: the pupil and iris point the way, and the amount of sclera revealed underlines it. Without a prominent sclera or a distinct pupil, the eyes of a dog don't indicate the direction of their attention nearly as much.

Closer still, and we begin to see serious species differences. Dogs manage to gather *more* light than we do. Once light enters the eye of a dog, it travels through the gel-like mass that holds nerve cells to the retina (we'll get to that in a moment), then through the retina to a triangle of tissue, which reflects it back. This *tapetum lucidum,* in Latin "carpet of light," accounts for all the photographs you have of your dog with brilliant light shining out where their eyes should be. Light entering the dog's eye hits the retina at least twice, resulting not in the redoubling of the image but in a redoubling of the light that makes images visible. This is part of the system enabling dogs to have such improved night and low-light vision. While we might make out a match being brightly struck in the distance on a dark night, the dog could detect the gentle flame on the lit candle. Arctic wolves spend a full half of the year living in utter darkness; if there is a flame on the horizon, they have the eyes to spot it.

EYES OF THE BALL-HOLDER

It is inside the eye—at that retina receiving light twice—that, one by one, characteristic habits of dogs can be traced to their anatomy. The retina, a sheet of cells on the back of the eyeball, translates light energy into electrical signals to our brains, which leads to our feeling that we've seen something. Much of what we see is given meaning only by our brains, of course—the retina just registers the light—but without the retina, we would experience only darkness. Even slight changes in the conformation of the retina can radically change vision.

There are two slight changes in the canine retinae: in the distribution of photoreceptor cells and in the speed with which they operate. The former leads to their ability to chase prey, retrieve a tossed tennis ball, to their indifference to most colors, and to their inability to see something right in front of their noses. The latter leads to their being uninterested in daytime soap operas left on for them when their owners are out of the house. We'll look at these in turn.

Go get the ball!

Some of the most important things for humans to see are any other humans situated within a few feet of their face. Our eyes face forward, and our retinae have *foveae:* central areas with an extra abundance of photoreceptors. Having so many cells in the center of the retina means that we are very good at seeing things right in front of us in high detail, great focus, and strong color. Perfect for identifying that blob of color and form coming at you as your boyfriend or your mortal enemy.

Only primates have foveae. By contrast, dogs have what is called an *area centralis:* a broad central region with fewer receptors than a fovea, but more than in peripheral parts of the eye. Things directly in front of a dog's face are visible to him, but they are not quite as sharply in focus as they would be for us. The lens of the eye, which adjusts its curvature to focus light onto the retina, doesn't accommodate to nearby sources of light. In fact, dogs might overlook small things right in front of their nose (within ten to fifteen inches), because they have fewer retinal cells committed to receiving light from that part of the visual world. You need no longer puzzle at your dog's inability to find the toy that he is nearly stepping on: he's not got the vision to take note of it until he takes a step back.

Breeds of dogs differ so much in their retinae that they see the world differently. The area centralis is most pronounced in those breeds with short noses. Pugs, for instance, have very strong areas centralis—almost fovea-like. But they lack a "visual streak," which dogs with long noses (and wolves) have. In Afghans and retrievers, for instance, the area centralis is less pronounced, and the photoreceptors of the retina are more dense along a horizontal band spanning the middle of the eye. The shorter the nose, the less visual streak; the longer the nose, the more visual streak. Dogs with the visual streaks have better panoramic, high-quality vision, and much more peripheral vision than humans. Dogs with the pronounced area centralis have better focus in front of their faces.

In a small but significant way, this difference explains some breed-based behavioral tendencies. Pugs are not typically so-called "ball dogs" but long-nosed Labrador retrievers are. Not because of their long noses per se. In addition to their ability to put their millions of olfactory cells to good use, Labradors are visually equipped to notice, say, a tennis ball traveling across the horizon, without having to shift their gaze. For a short-nosed dog (as for all humans of any nose length), a ball tossed to the side just disappears into the periphery if they don't follow it with their head. Instead pugs are probably better at bringing close objects into focus—say, the faces of their owners on whose laps they sit. Some researchers speculate that this relatively blinkered vision makes them more attentive to our expressions, and seem more companionable.

Go get the green ball!

Dogs are not color-blind, as is popularly believed. But color plays a much less important role for them than it does for humans, and their retinae are why. Humans have three kinds of cones, the photoreceptors responsible for our perception of details and of colors: each fires to red, blue, or green wavelengths. Dogs have only two: one is sensitive to blue and the other to greenish-yellow. And they have fewer of even those two than humans do. So dogs experience a color most strongly when it is in the range of blue or green. Ah, but a well-scrubbed backyard pool must seem radiant to a dog.

As a result of this difference in cone cells, any light that looks to us like yellow, red, or orange simply doesn't look the same way to a dog. Consequently, they seem perfectly oblivious when you ask them to bring back grapefruits from the store and grow irritated when they bring back tangerines. Still, orange, red, and yellow objects might still look different to them: the colors have different brightnesses. Red may be seen by them as a faint green; yellow a stronger one. If they seem to be able to discriminate red and yellow, they are noticing a difference in the *amount* of light these colors reflect toward them.

To imagine what this might be like, consider the time of day when our color system breaks down: in the dusk right before night. If you're outside in a park, in your yard, or anywhere nature lives, take a look around you. You might notice the wash of exuberant green leaves above you subtly dulling to a more unassuming hue. You can still see the ground underfoot, but details—the distinctness of blades of grass, the layers of petals— are reduced. Depth of field squashes somewhat. I tend to stumble more than usual on protuberant gray rocks that blend with the earth. The reason for this loss of visual information is anatomical. Cones, clustered toward the center of the retina, are not sensitive to low light, so they don't fire as often at dusk or night. As a result, our brains get signals from fewer cells detecting colors. And the near world flattens out a little: we can still see that there is color, and we still detect lights and darks, but the richness of colors has fallen away; colors are grainier, less detailed. So it might be for dogs, even at noontime.

As they do not experience a great range of distinct colors, dogs rarely show color preferences. Your clashing choice of red leash and blue collar affects your dog not at all. But a deeply saturated color may get more attention from a dog, as will an object placed in a background of contrasting colors. It may be meaningful that your dog attacks and pops all the blue or red balloons left over after the birthday party winds down: they are most distinct among a sea of pastels.

Go get the green bouncing ball . . . on the TV!

Dogs make up for their dearth of cones with a battery of rods, the other kind of photoreceptor in the retina. Rods fire most in low-light situations and at changes in light densities, which is seen as motion. In human eyes, rods cluster at the periphery, helping us notice something moving out of the corner of our vision, or when the cones slow their firing at dusk or night. The density of rods in dogs' eyes varies, but they have as much as three times as many rods as we do. You can make that ball your dog is not seeing directly in front of him magically appear by giving it a little shove. Acuity greatly improves for close objects when they are bouncing.

All these differences in the dog's perception, experience, and behavior result from some small changes in the distribution of cells on the back of the canine eyeball. And there is another small change that results in a large difference—potentially more far-reaching than a change in focal area or color vision. In all mammalian eyes, rods and cones make electrical activity out of light waves by means of a change in the pigment in the cells. The change takes time—a very small amount of time. But in that time, a cell processing light from the world cannot receive more light to process. The rate at which the cells do this leads to what is called the "flicker-fusion" rate: the number of snapshots of the world that the eyes take in every second.

For the most part, we experience the world as smoothly unfolding, not as a series of sixty still images every second, which is our flicker-fusion rate. Given the pace at which events that matter to us happen, this is usually plenty fast. A closing door can be grabbed before it slams; a handshake received before it is withdrawn in annoyance. To create a simulacrum of reality, films—literally "moving pictures"—must exceed our flicker-fusion rate only slightly. If they do, we do not notice that they are just a series of static pictures projected in sequence. But we *will* notice if an old-fashioned (pre-digital) film reel slows down in the projector. While ordinarily the images are being shown to us faster than we can process them, when it slows we see the film flickering, with dark gaps between the frames.

Similarly, fluorescent lights are so annoying because they operate too close to the human flicker-fusion rate. The electrical devices used to regulate the current in the light function right at sixty cycles per second, which those of us with slightly faster flicker-fusion rates can thus see as a flicker (and hear as a buzz). All indoor lights fluorescently flicker to houseflies, with extremely different eyes than ours.

Dogs also have a higher flicker-fusion rate than humans do: seventy or even eighty cycles per second. This provides an indication why dogs have not taken up a particular foible of persons: our constant gawking at the television screen. Like film, the image on your (non-digital) TV is really a sequence of still shots sent quickly enough to fool our eyes into seeing a continuous stream. But it's not fast enough for dog vision. They see the individual frames and the dark space between them too, as though stroboscopically. This—and the lack of concurrent odors wafting out of the television—might explain why most dogs cannot be planted in front of the television to engage them. It doesn't look real.*

One could say that dogs see the world faster than we do, but what they really do is see just a bit *more* world in every second. We marvel at dogs' seemingly magical skill at catching a Frisbee on the fly, or following a rapidly bouncing ball. Their Frisbee-catching procedure, as has been documented with microvideo recording and trajectory analysis, turns out to match nicely the navigational strategy naturally used by baseball outfielders to line themselves up with the arc of an incoming ball. Excepting a few phenomenal outfielders, dogs actually see the Frisbee's, or

*The conversion to entirely digital television broadcasts will eliminate the flicker-fusion problem, making TV-viewing more viable (but no more olfactorily interesting) for dogs—who are no doubt ambivalent.

the ball's, new location a fraction of a second before we do. Our eyes are internally blinking in those milliseconds that a flung Frisbee is moving along its course toward our heads.

Neuroscientists have identified an unusual brain disorder in some humans called "akinetopsia." These akinetopsics have a kind of motion blindness: they have difficulty integrating a sequence of images into the normal perception of motion. A person with akinetopsia may begin pouring a cup of tea and then not register a change until many images later, by which time the cup is overflowing. As non-brain-damaged persons are to akinetopsics, dogs are to us: they see the interstices between our moments. We must always seem a little slow. Our responses to the world are a split second behind the dogs'.

VISUAL UMWELT

With age, Pump suddenly became reluctant to enter the elevator, perhaps not seeing it well in the darkness after being outside. I encourage her, or jump in myself first, or throw something light-colored on the elevator floor for her to see. Finally, every time, she rallies and leaps in, as though crossing a great crevasse, brave girl.

So dogs can see some of the same things we do, but they don't see in the way that we do. The very construction of their visual capacity explains a broad swath of dog behavior. First, with a wide visual field, they see what is around them well, but what is right in front of them less well. Their own paws are probably not in terrific focus to dogs. What wonder then how little they use their paws, relative to our reliance on the end of our forelimbs, to manipulate the world. A small change in vision leads to less reaching, grabbing, and handling.

Similarly, dogs can bring our faces into focus, but detect eyes less well. This means they will catch your full facial expression better than a meaningful glare, and they will follow a point or a turn better than a surreptitious glance out of the corner of the eye. Their vision complements their other senses. While they can locate a sound in space only roughly, their hearing is good enough for them to turn their eyes in the right direction, so they can search further visually . . . and then examine closely by nose.

For instance, dogs recognize us by our smells—but they also clearly look at us. What are they seeing? If your smell is not available—you are downwind or you're covered in perfume— they can use visual cues exclusively. They will hesitate if they hear your voice calling them, but it is not your face atop the approaching person, or your particular walk, or your mouth moving to call their name. Recent research confirmed this by examining dogs' behavior when they heard their owner's voice or a stranger's voice, accompanied by either a picture of the owner's face (on a large monitor) or of a stranger's face. The dogs looked longer at the incongruous faces: the owner's face, when paired with a stranger's voice, and the stranger's face, when it appeared with the owner's voice. If it were just that the dog preferred the owner's face, they would have always gazed at that face the longest. Instead they looked longest when there was something surprising: a mismatch.

The physical elements of vision define and circumscribe what the dog experiences. There is a further element of that experience: the role vision plays in the hierarchy of senses. For visual creatures like humans, there is particular delight when we encounter something through one of our non-visual senses first. To arrive outside my apartment door and smell something wonderful—to open the door and hear the sounds of sizzling in pots, the clank of silverware; to be instructed to taste a forkful of the pot's contents with my eyes closed—renders a familiar experience new. I only come forth to verify the scene with my eyes: my boyfriend in front of me with dinner in messy preparation around him.

Coming to something through the secondary senses first discombobulates, then introduces a feeling of novelty to the ordinary. As dogs have their own hierarchy of senses, I imagine that they too might feel the mystery of coming at something by means other than their nose. This may explain both the difficulty dogs have in understanding some of our first requests to them (*off the couch!* I said to my new puppy, as she looked at me searchingly), and the pride they seem to take in learning a distinction from our visual world.

Though our visual worlds overlap, dogs attach different meanings to the objects seen. A Seeing Eye dog must be taught the umwelt of the human: the objects that are important to the blind person, not those of interest to the dog. Try yourself getting your dog to even acknowledge the existence of a sidewalk curb. What is a curb to a dog? With persistence, dogs can be taught, but most dogs simply do not *see* a curb: it is not that the curb is invisible, but that it lacks any important meaning to them. The surface below their feet may be rough or soft, slippery or rocky, it may hold the scent of dogs or of men; but the distinction between the sidewalk and the street is a human distinction. A curb is but a slight variation in altitude of the hardened mass with which we cover the dirt, which only has a meaning to those who concern themselves with such concepts as *roads, pedestrians,* and *traffic.* The guide dog must learn the importance of the curb to his companion. He must learn the significance of a speeding car, a mailbox, other people approaching, a doorknob. And he will: he may begin to associate the curb with the distinctive striping of a crosswalk, with the dark, smelly rain gutters that run along them, or with the change in brightness from the concrete to the asphalt. Dogs are much better at learning about things that are important to us in our visual world than we seem to be in understanding theirs. I still can't tell you why Pump became excited at the mere sight of a husky-shaped dog appearing around the corner. But after a dozen years I began to notice that she did. She, on the other hand, was quicker to recognize the importance I placed on certain objects—the distinction between the frayed sofa and my favored armchair with respect to her chance of sitting on it; the slippers whose fetching made me laugh versus the running shoes whose delivery made me scold.

There is a final, unexpected facet of the visual experience of the dog: they see details that we cannot. The fact of dogs' relatively weak visual capacity turns out to be a boon to them. Since they are not trying to take in the whole world with their eyes alone, they may see details that we don't notice. Humans are gestalt lookers: every time we enter a room, we take it all in in broad strokes: if everything is more or less where we expect it to be . . . yes . . . we stop looking. We don't examine the scene for small, or even radical changes; we might miss a gaping hole in the wall. Don't believe it? At every moment of our lives we are not noticing a gaping hole: one in our visual field caused by the very construction of our eyes. The optic nerve, the neural route conveying information from the retinal cells to brain cells, tunnels right through the retina on its way back to the brain. Thus if we hold our eyes still, there is a part of the scene in front of us that is not captured on our retina—as there is no retina there to capture it. It's a blind spot.

We never notice this gaping hole in front of us because our imaginations fill in that spot with what we expect to be there. Our eyes dart back and forth constantly and unconsciously— movement called *saccades*—to further complete the visual scene. We never experience the missing spot. In the same way, we also have a blind spot for those things that are slightly different—but close enough—to what we expect to see. As well-adapted visual creatures, our brains are equipped to find the sense in the visual information sent it, despite holes and incomplete information.

We are maybe too well adapted. Some of what we overlook, animals see. The celebrated autistic scientist Temple Grandin has demonstrated the reality of this with cows, for instance. Often cows being led along wending chutes into the slaughterhouse balk, kick, and refuse to proceed. As far as we know, this is not because they understand what will happen in the slaughter- house. Instead there were small visual details that surprised or frightened them. The reflection of light in a puddle; an iso- lated yellow raincoat; a sudden shadow; a flag flapping in the breeze: seemingly insignificant details. We are certainly able to see these visual elements—but we do not notice them as cows do.

Dogs are closer to those cows than to us. Humans quickly label and categorize a scene. Walking to work along a Manhat- tan street, the typical commuter is perfectly oblivious to the world he is passing. He notices neither beggars nor celebrities; startles neither to ambulances or parades; simply sidesteps a crowd gathered to gape at . . . well, whatever it is crowds gape at: I rarely stop to see. On most mornings, the route is reduced to its landmarks; nothing else needs attending to. There is good reason to believe that this is not how dogs think. The walk to the park becomes familiar over time, but they don't stop looking. They are much more struck by what they actually see, the immediate details, than what they expect to see.

Given how dogs see, how do they apply their visual ability? Cleverly: they look at us. Once a dog has opened up his eyes to us, a remarkable thing happens. He starts gazing at us. Dogs see us, but the differences in their vision also seem to allow them to see things about us that even we do not see. Soon it seems they are looking straight into our minds.

Seen by a Dog

I am startled and a little flustered to look up from my work and see Pump watching me, her eyes trained on mine. There is a powerful pull to a dog who looks you in the eyes. I am on her radar: it feels that she is looking not just at me, but to—and into—me.

Look a dog in the eyes and you get the definite feeling that he is looking back. Dogs return our gaze. Their look is more than just setting eyes on us; they are looking at us in the same way that we look at them. The importance of the dog's gaze, when it is directed at our faces, is that gaze implies a frame of mind. It implies attention. A gazer is both paying attention to you and, possibly, paying attention to your own attention.

At its most basic level, *attention* is a process of bringing forward some aspects of all the stimuli bombarding an individual in a moment. Visual attention begins with *looking;* auditory attention with *hearing:* both are possible for all animals with eyes and ears. Just having the sensory apparatuses isn't sufficient to do what we generally mean by *paying attention,* though: considering what it is one has turned to look or hear.

When invoked by psychologists, attention is treated not just as turning the head toward a stimulus, but as something else in addition: a state of mind that indicates interest, intent. In attending to someone else's head turns, one may be demonstrating an understanding of the psychological states of other people—a distinctively human skill. We attend to others' attention because it helps to predict what that someone other will do next, or what he can see and what he might know. One of the deficits that many people with autism have is an inability, or lack of inclination, to look at other people's eyes. As a result, they aren't instinctively able to understand when other people are paying attention—or how to manipulate others' attention.

The simple ability to focus on some things while ignoring others is crucial for any animal: objects one sees, smells, or hears may be more or less relevant for survival. Attend to those that are relevant; ignore the rest of the visual landscape or the confusion of sounds. Even with survival no longer our most pressing concern, humans are constantly trying to direct, divert, or attract attention. Some attention mechanism is required to do all the ordinary things of our days: to listen to someone talking to us, to plan a walking route to work, even to remember what one was thinking a moment ago.

Dogs, social animals like us, and also more or less relieved of survival pressures, surely have some interesting mechanisms with which to attend to the world. By virtue of their different sensory abilities, though, they are able to attend to things we never notice, such as how our odor changes through the day. Likewise, we focus carefully on things that dogs do not even detect, such as subtle differences in language use.

But what distinguishes dogs from other mammals, even other domesticated mammals, is the way that their attention overlaps with ours. Like us, they pay attention to *humans:* to our location, subtle movements, moods, and, most avidly, to our faces. A popular conception of animals is that if they look at us at all, it is from fear or appetite, monitoring us as possible predator or prey. Not true: the dog looks very particularly at humans.

Just how particularly is the subject of a mad rush of contemporary research into dog cognitive abilities. This research uses as markers the landmarks in the development of human infants into human adults, which is well documented, and which result is obvious: by adulthood, we all understand what it means to pay attention. What the dog research is revealing is that dogs have some of the same abilities that we do.

THE EYES OF A CHILD

For dogs and humans both, it all begins with a few innate behavioral tendencies. Having and understanding attention is not automatic, but it develops naturally from these instincts. Human infants, like most animals, have a basic orienting reflex: move, as best or as much as you can, toward a source of warmth, food, or safety. Newborns turn their faces toward warmth and suck: the rooting reflex. At that age, infants can do little more. Ducklings, more precocious, relentlessly chase after the first adult creature they see.* In both ducklings and humanlings, this reflex relies on an early perceptual ability: having at least *noticed* the presence of others. It is an ability that helps us, in our first few years, learn about the important fact of others' attention.

---

*Ethologist Konrad Lorenz beautifully demonstrated this tendency of young waterfowl in the 1930s by positioning himself as that first adult creature seen by a gaggle of greylag goslings. They followed him readily, and Lorenz wound up raising the brood as his own.

For humans, there is a reliable course of development through infancy of certain behaviors associated with this growing understanding of other people. It is all about learning to attend to the right—human—things in the world, and beginning to understand that others are attending, too. And it begins as soon as they open their eyes. Newborn babies can see, although not much. They are incredibly nearsighted: peering, cooing faces brought just inches from their own may be clear, but that is about the extent of clarity of the world. One of the first things infants notice is any faces nearby. In fact our brains have specialized neurons that fire when we see a face. Infants can detect, and prefer to look at, a face or something facelike—even three points forming a V—rather than other visual scenes. From early in their lives, infants stare longer* at that which interests them, the mother's face being among the first items of interest. Soon infants also learn to distinguish a face looking toward them from one looking away. This is a simple skill, but not a trivial one: out of the visual cacophony of the world, they must start noticing that there are objects, that some of those objects are alive, that some of those alive objects are of particular interest, and that some of those interesting live objects attend to you when they face you.

Once that is established, and their own visual acuity improves, infants focus on the details in that face. They delight at peekaboo: a game playing simply with the importance of eyes. As psychologists have shown by sticking out their tongues and making faces at infants, very young infants can imitate sim-

---

*Developmental psychologists rely on the fact that though infants cannot report what they are thinking, they reliably look longer at things that interest them. By using this one feature of infant behavior, psychologists gather data about what the infant can see, distinguish, and understand and what he prefers.

ple expressions. Of course, these expressions don't have the meaning that they will later (we must assume that the infant is not actually sticking his tongue out spitefully at the psychologist, though one might wish it so). Infants are simply learning to use their facial muscles. By three months, they've got it, and they start reacting to others by making faces and smiling socially. They move their heads to look at other faces nearby. By nine months old they are tracking other people's gaze and seeing where it lands. They might use that gaze to find some object that they have asked about, or that has been hidden from them. Soon they extend the line of gaze into a point with finger, fist, or arm to request an object, and by their first birthday, to show or share.

These behaviors reflect the infant's burgeoning understanding that other people have attention, attention that can light on objects of interest: a bottle, a toy, or them. Between twelve and eighteen months, they begin to engage in bouts of *joint attention* with others: locking eyes, then looking to another object, then back to eye contact. This marks a breakthrough: to achieve full "jointness" the infant must on some level understand that not only are they both looking together, they are *attending* together. They are understanding that there is some invisible but real connection between other people and the objects that are in their line of vision. Once they do this, all hell can break loose. Infants can start manipulating others' attention simply by gazing someplace. They check where other people are looking and pointing, and they begin to notice if adults are looking at them while they are doing activities they want to share (or conceal). They will give an anticipatory glance at an adult before pointing or showing themselves. They work very hard to get attention looking at them. And they may begin avoiding attention: going out of a room at key moments, or concealing objects from an adult's view. (This prepares them well for being difficult adolescents.)

We all become characteristically human by this same developmental route. Within a few years an infant goes from aimlessly looking out of new eyes, to looking meaningfully, to gazing at others, to following the gaze of others. They happily hold eye contact. Before long they are using gaze to get information, to manipulate the gaze of others—by distraction, gaze avoidance, or pointing—and to get attention. At some point, they come to a realization about the fact of the mind behind someone else's gaze.

THE ATTENTION OF ANIMALS

> She comes within an inch of me and starts panting at me, eyes wide and unblinking, to tell me that she needs something.

Step-by-step, cognition researchers have been tracing this developmental course with a new subject: non-human animals. How much of the infant's trajectory is followed by animals? After they open their eyes, do they look with intention? Do they notice others' eyes? Do they understand the importance of attention?

This is one facet of the study of animal cognition, which asks what an animal subject understands about the "mental states" of others. Most of the experimental tests run with animals are of the kind we feel sure we humans excel at: tests of physical and social cognition. Captive animals from sea slugs to pigeons to prairie dogs to chimpanzees have been set into mazes; presented with counting, categorizing, and naming tasks; asked to discriminate, learn, and remember series of numbers and pictures. Tasks are devised to see if they recognize, imitate, or deceive others—or even recognize themselves. And in some tests, the question is even more characteristically human: of the kind of social thinking going on when animals interact—with members of their own species, and with those of other species. When a caged chimpanzee looks at a human attendant, is he considering anything about the attendant? Does he wonder how to get her to open the door (does he wonder anything at all?), or is he simply waiting to see what this colorful, animate object nearby does that might be relevant or interesting? Does a cat consider that mouse as an agent, as an animal with a life—or does he see the mouse as a moving meal that must be stopped and dismantled?

As we've touched on already, the subjective experience of animals is notoriously difficult to get at scientifically. No animal can be asked to relate its experience in voice or on paper,* so behavior must be our guide. Behavior has its pitfalls, too, since we cannot be positive that any two individuals' similar behavior indicates similar psychological states. For instance, I smile when I am happy . . . but I may also smile out of concern, uncertainty, or surprise. You smile back at me: it too might be happiness—or ironic detachment. To say nothing of the near impossibility of determining whether your "happiness" feels like mine does.

Still, even without having constant verification of others' mental states, behavior is a good enough guide to allow us to predict an animal's future behavior well enough to interact

---

*Well, for the most part: Kanzi the bonobo and Alex the African gray parrot are among those who have been asked and have answered: Alex was able to create and utter novel, coherent, three-word sentences based on a vocabulary built from eavesdropping on researchers; Kanzi has a multihundred-word vocabulary of lexigrams (symbolic pictures) that he can point at to communicate. And a single dog, Sofia, has been trained to use a simple eight-key keyboard concurrent with events she had already learned, like going for a walk, going into a crate, and getting food or a toy. She learned to press the appropriate key to make a request. As a communication, this behavior is closer to asking for dinner by bringing an owner an empty food bowl than it is to a full-fledged language. More abstract utterances have not been reported (nor abstract keyboards designed).

peacefully and productively. Thus we study what animals do—in particular, what they do that is like what humans do. Since using and following attention is so important in human social interaction, animal cognition researchers look for behaviors that indicate that an animal is using attention.

Dogs have recently trotted gamely into experimental labs, controlled outdoor facilities, and onto data sheets meant to gather information about their abilities at using attention. The dogs are put in controlled settings, usually with one or more experimenters present, and a hidden, desirable object: a toy or a food treat. By varying the cues that they use to inform the dogs about the location of the treat, the experimenters aim to determine which ones are meaningful to the dogs.

The question for researchers is just how far along these stages of the child's development of attention dogs go. Attention begins with gaze, and gaze requires visual capacity. We have already established what dogs can see; we know that they look. Do they understand attention?

Mutual gaze

A gaze is more than it seems: by gazing at someone, one very nearly *acts* upon him. As my students discover in their field experiments, eye contact comes close to feeling like actual tactile contact. There are undiscussed and yet widely shared rules governing eye contact with others—violation of which may be seen as an act of aggression or of intimacy. We may stare down someone in an attempt to subdue them, or, alternatively, use a long, steady gaze to indicate a more lustful interest.

With a little variation, this could as easily describe how many non-human animals use eye contact. Between apes, eye contact is steeped with importance: it can be used as an aggressive action, and will be avoided by a submissive member of a troop. To stare at a dominant animal is to invite yourself to be attacked. Not only do chimpanzees avoid staring, they avoid being stared at. Subordinate chimps carry themselves despondently, looking down at the ground or their own feet and only furtively glancing around them. In wolves, too, a direct stare may be taken as a threat. So the "aggressive" element of eye contact is the same as with humans. The variation is this: all non-human animals with any meaningful visual capacities will turn their eyes to something of interest—but if the thing-of-interest is a member of their species, the social pressure of gazing usually deflects the gaze of interest.

Thus we can expect that dogs might act somewhat differently than we do with regards to mutual gaze. As dogs evolved from a species in which a stare is most often a threat, we might do best to consider their avoidance of eye contact less an inability than a result of their evolutionary history. But wait! Dogs do look at our faces. They look at each other in the center of the face: at eye level. Most dog owners will report that their dogs gaze at them directly in the eyes.*

---

*One could make an argument that this behavior was reinforced because of the survival value of looking at humans. As with infants, an adult face will hold much information, not the least of which could be where the next meal is coming from. The early-twentieth-century ethologist Niko Tinbergen similarly found that baby gulls have a strong attraction to the red-dotted beaks of adult gulls (and to any stick with a red dot placed on it by an ethologist, too).

So something changed with dogs. While the threat of aggression prevents mutual gaze among wolves, chimpanzees, and monkeys, for dogs the information to be gained by looking us in the eyes is worth enduring any residual, ancient fear that a stare might cause an attack. That humans respond well to a dog gazing at them is a happy circumstance—and our bond with them is thereby strengthened.

To be sure, it may be less "eye contact" than "face contact."* Because of the superficial anatomy of the dog eye—the lack of distinct iris and whites of the eye—specific eye direction can often only be confirmed from closer range than scientists' video cameras have gotten. Generations of dog breeders have tended to prefer the trait of dark eyes in their charges. Dogs with light-colored irises are often thought to look volatile or sneaky—ironically because we can see clearly when they avoid eye contact. By breeding out the light irises we do not eliminate shiftiness, just our *awareness* of the fact that dogs shift their gaze. Darting eyes become less conspicuous. We sleep better at night with a calm-featured dog at the foot of our bed than with a nervous eye-darter. For all intents and purposes, though, we can say that a dog and human "mutually gaze" when we turn our faces toward each other.

The primal pull of gaze still affects dogs' behavior. If you stare unblinkingly at your dog, he may look away. Approached

*Dogs show an additional tendency, one that people do, too, when looking at faces: to look leftward first (that is, to the right side of the face). Even young children show this "gaze bias": looking first, and longer, to the right side of an examined face. By closely observing dogs observing faces, researchers have found that dogs share this bias—when looking at human faces. When looking at other dogs, they show no gaze bias at all. Why this might be is still a matter of conjecture: perhaps we express emotions differently on each hemi-face; and perhaps dogs emote more symmetrically (lopsided ears aside). Dogs have learned to look at humans the way humans look at humans.

by a dog who appears overly aggressive or overly interested, a dog can diffuse some of that excitement by glancing to the side. Your chastisement or accusation of your dog accompanied by a glare may also provoke a demure averral of the dog's gaze. Given the easily recognized shifty look of the guilty man confronted by his accuser, it is no wonder that we attribute the same to the gaze-avoiding dog. The refusal to look us in the eyes contributes to a look of guilt—especially when we are already certain they have done something to inspire it. Whether they are themselves feeling guilt or atavism is not obvious.

But the fact that dogs will look us in the eyes allows us to treat them as a little more human. We apply to them the implicit rules that accompany human conversations. It is not uncommon to see a dog owner pause from scolding a "bad dog" to physically turn the dog's head back toward the owner's face. We want dogs to look at us when we are talking to them—just as we use gaze in human conversation, in which listeners look at the face of the speaker more than the reverse. (Notably, we do not stare at each other non-stop in conversation, and might feel unsettled if someone did.) There is more direct eye contact among humans speaking intimately or honestly, and we tend to extend that conversational dynamic to our dogs. We call their names before speaking to them, treating them like willing, if taciturn, interlocutors.

Gaze following

It doesn't happen at once, but not long after bringing a dog or puppy to your home for the first time you might notice something: nothing in the house is safe. Dogs train humans to become suddenly tidy: putting away shoes and socks almost as soon as they are removed; taking out the trash well before it heaps high; leaving nothing on the floor that could fit into the gaping mouth of a teething, excited, unrestrained pup. A temporary peace may ensue. After all, you can put things behind closed doors, into shut cabinets, and onto high shelves. Dogs look, baffled, in the area from which the (shoe, takeout container, hat) has mysteriously gone missing. But soon you will notice that the dog has learned something new: *you* are the source of the mysterious relocations—and you have a tendency to tip your hand.

How? You look. When we pick up the sock and set it down, we're not just attached to it by a hand; the action is accompanied by a gaze. We look where we are going. Later, we may look again at that safe sock spot when discussing the dog's earlier thievery. Again our gaze reveals the location of the sock: the gaze is itself information. We have already met this ability to use the direction of another's gaze, so-called *gaze following,* which infants do before reaching one year old. Dogs do it even sooner.

A gaze that intends to share information is simply a point done without hands. Following a point is a slightly simpler ability. Certainly dogs see a lot of pointing and gesturing as they observe human members of their families. This may be the source of their gaze-following ability, or it may simply bring out an innate ability to glean any information possible from our behavior. Researchers have tested the limits of their ability, natural or learned, in various experiments that put dogs in a context where they can get information from a person's pointing gesture. For instance, a biscuit or other desired food might be hidden under one of two inverted buckets while the dog subject is out of the room. When all odor cues are masked, the dog has to make a decision which bucket to choose. If he chooses correctly, he is rewarded with the food; if not, he is rewarded with nothing. A person who knows which bucket to choose is standing nearby.

Chimpanzees have been given variations of this task in captive research settings. Surprisingly, though they seem to follow points, they don't always do well at following gaze alone.

Dogs perform admirably. They follow points, points that reach across the pointer's body, points from behind the body, and are even better if the point includes a finger further signifying the baited bucket.* They haven't simply learned the importance of an outstretched arm. Pointing with elbows, knees, and legs also serves as information. Given even a momentary point—a glance of a point—the information is theirs. They can follow the pointed cue given by a life-sized video projection of their owner. Though they have no arms with which to point themselves, they outperform the chimps who have been tested. Best, dogs can use simply the person's head direction—her gaze—to get information. You may be able to hide that sock from your sock-coveting chimpanzee, but your dog will spot it.

Where dogs' use of attention really gets interesting is in less overt cases. Not just when we point and they look, but when they have to decide how to inform us that they need to go outside—

---

*It should be noted that this skill is affirmed by dogs' following a pointing hand to one of two baited buckets at rates "significantly above chance." What this means is that they don't choose which bucket to search under first randomly. Instead they choose the bucket pointed at from 70 to 85 percent of the time. That's good, but they are still making the wrong judgment 15 to 30 percent of the time! Three-year-old children get the right bucket every single time. What this indicates is that the dogs' success is probably the result of a mode of understanding that is not identical to ours.

or want a ball tossed to them. Or they need to tell us some very important news about where a tasty treat fell out of their reach while we were out of the room. Play with humans is a rich context for the possible appearance of some of these abilities; experimental paradigms also manipulate the information that can be gleaned from others' attention. All signs indicate that dogs seem to understand how to get attention, how to make requests of us using attention, and what kind of *in*attention allows them to get away with bad behavior.

Attention-getting

The first of these abilities, when seen in children, is called "attention-getting." Informally, you may know it as anything that your dog does to interfere with what you are currently trying to do. More formally, these are behaviors that are sufficient to change the focus of someone else's attention, by stepping into his visual field, making a discernible noise, or making contact. Suddenly jumping on you is a familiar dog attention-getting behavior, if not one that is well loved by the jumped-upon. Barking is another. Their attention-getting means aren't restricted to the quotidian, though. Less recognized means include bumping, pawing, or simply orienting oneself right up in front of someone else: what I have called an *in-your-face* in my data of dog play behaviors. Guide dogs use "sonorous mouth licking"—audible slurps—to get the attention of their visually impaired charges when needed. The excitement of play sometimes leads them to come up with novel techniques, too. My favorite sessions to observe are those in which an eager but frustrated dog mirrors the behavior of the object of his unrequited play interest: approaching and drinking from the bowl from which the first dog is drinking—and using it as a means toward licking his face; or grabbing a stick of his own when another dog finds a good stick to be sufficient company.

Dogs use attention-getters regularly with us, and are often rewarded with our attention. But unless they show some subtlety in application of these behaviors, their use does not prove their full understanding of our attention. It may be that they are simply throwing all the tools they have at the problem of needing you to look at them. A child hollers, you come racing to his side: an attention-getter is born. Observations of dogs playing with humans show just how crude or subtle their use of these behaviors is. There are dogs who will stand barking continuously over a retrieved tennis ball, while their owners socialize with members of their own species. While a good attention-getter, a bark is not well applied if it continues to be used after it has failed to get attention. On the other hand, there is also evidence of very subtle visual attention-getters by dogs in reaction to the divided attention of their owners. By changing posture, as from a seated posture to standing, or from standing to approaching, dogs are able to reengage the owner enough to toss a ball or lunge playfully.

You are regularly a witness to the flexibility dogs show in attention-getting. If your dog did not rouse you from your armchair and novel by simply approaching you, he may have wandered off only to return carrying a shoe or another verboten item. Probably this causes you to chastise him gently and return to your book. More serious tactics are needed, he sees. Next may come whining, or a tentative woof; a tactile intervention—a slight push with a wet nose, nuzzling, or jumping; even a loud drop to the floor at your feet with a sigh. They are trying their best with you here.

Showing

So far, the dogs have kept pace with the developing child: gazing, following a point, following gaze, and using attention-getters. Do they also point, as best they can, with their bodies? Do they point with their heads to *show* you something?

Here again, experimenters set up a situation that they presumed would prompt the behavior, if the ability exists. The scenario is the gaze-following task, inverted. Instead of being the naïve ones, in these cases the dogs are informed but impotent: they alone witness an experimenter hiding a treat, damnably out of their reach. Their owners then wander into the room, and experimenters train their cameras on the dogs: Do they see the owners as tools which can help them? If so, do they communicate the location of the treat?

In these cases, it seems the only obtuse animals in the room are the humans, who may not see the dog's behavior as potentially showing them anything. That behavior consists of a lot of attention-getters (such as barking), followed, critically, by looking back and forth between the owner and the location of the treat. In other words, pointing with that gaze: showing.

This is visible daily in non-experimental settings. Ball dogs crazy about retrieving generally deliver their slobbered spheres to the front side—the face—of the ball tosser, not to her back. And, if the ball is mistakenly dropped at the unresponding backside of the owner, the dog has an arsenal of attention-getters to employ, followed by relentless gaze alterations—looking at the face of the human holding the ball, and looking back at the ball in quick succession. The restless, attention-starved dog is never satisfied dropping found socks at your back; they are left within your sight, if not right on your lap.

Manipulating attention

Finally, dogs use the attention of others as information, both to get something they want and, more remarkably, to determine when they can get away with something.

Research has determined this by asking if dogs choose intelligently when given a choice about whom to request food from. If every person is an equally good source of food, one would expect that dogs would approach all persons with that same beguiling expression—half entreaty, half expectation. There are dogs who do so, of course,* and those who reserve their begging for butchers, or owners who stuff their pockets with liver treats. But most dogs make a distinction that is important to us when we desire something: between possible and impossible collaborators. We make requests appropriate to the state of knowledge and capacity of our audience. You do not ask the baker to explain string theory and the physicist for a loaf of seven-grain, sliced.

In experimental settings mining the same four elements of dog, experimenter, food, and knowledge, dogs seem to distinguish between humans who might be helpful to them and humans who will likely not. When a person with a sandwich is either blindfolded or facing away, dogs suppress the urge to stay as close as possible to the sandwich. Instead, if there is a non-blindfolded person nearby, they go beg to him instead. Let this be a lesson that begging at the table is probably encouraged by your eye contact toward the dog—even just long enough to tell him *no begging!* Alternately, set up one person as the responsive,

---

*Dogs who often wind up being called *people dogs* for their keen interest in owners over dogs.

looking beggee and all the dog's attention will go to him. (Children are good for this role.)

Dogs also approach the blindfolded persons warily—as befits the situation, if one isn't let in on the fact of being a subject in an experiment. These experiments using unresponsive, oddly outfitted characters are typical of psychological tests. At some level, they are useful in order to avoid the possibility that the subject has had experience with the setting they are about to encounter. In other words, the tests aim to get at what the dogs intuitively understand about the knowledge states of the human, not what the dog might have learned about what to do when you see someone who is blindfolded. Still, the dog is confronted with what must be a strange couple of hours.

Variations of the begging trials were first run with chimpanzees. In that context, the attentional state of the human was taken to indicate something about her knowledge. Someone who sees food baited in one of two hidden bins is "knowledgeable"; someone who stands idly by in the same room, but has a bucket over her head, is not. Did the chimps then beg to the knowledgeable person or to the one who is guessing at the location of the food (by chance guessing correctly once in a while)? Over time, chimps learn to beg to the knowledgeable informant—but only when the guesser has been out of the room, or has her back turned when the bin is baited. When the guesser simply has her eyes blocked—with a bucket, paper bag, or blindfold—the chimps begged to her, too.

Dogs have gone through trials with odd humans wearing buckets, blindfolds, or holding books in front of their eyes, blocking their vision. They outperform chimps: dogs preferentially beg to the looking—to those whose eyes they can see. This is just how we act, preferring to talk, cajole, invite, or solicit those whose eyes are visible. Eyes equal attention equals knowledge.

Best, dogs use this knowledge for manipulative ends. Researchers have found that dogs not only understand when we are attentive, but are sensitive to what they can get away with at different levels of their owners' attention. In one experiment, after being instructed to *lie down* (and dutifully so doing), dogs were observed in three trials. In the first condition, an owner stood and stared at her dog. The result? The dog stayed lying down: perfectly obedient. In the second condition, the owner proceeded to sit down and watch television: here the dog paused, but shortly disobeyed and got up. And in the third condition, the owner didn't just ignore the dog but left the room entirely, leaving the dog alone with his owner's command still echoing in his ears.

Apparently the echo was not long-lasting, for in these trials the dogs were quickest and likeliest to disobey the same command so well heeded when the owner was around. What is surprising is not that the dogs disobeyed when the owner left. It is, instead, that dogs do what two-year-olds, chimps, monkeys, and no other animals seem to do: simply notice exactly how attentive someone is, and vary their own behavior accordingly. The dogs methodically used the level of their owners' attention to determine under what circumstances they were free to break the owners' rules—just as they used the information from other dogs to get attention back toward them in play.

The dogs' attention-reading is highly contextual, however. When the same experiment was run using food, that great motivator to perform at their best, the threshold to disobedience was lowered: dogs disobeyed more quickly, and at lower levels of owner distraction. When the owner's attention was harder to gauge—when she was talking with someone else, or sitting quietly with closed eyes—the dogs' behavior was mixed. Some sat patiently, but, seemingly gathering steam, were prepared to spring up at once as soon as the owner left the room. Other dogs took even longer to disobey when the owners left the room than when they were in it but otherwise engaged. This illogic might be explained by a developmental fact, one that would vary dog to dog. Some owners establish a routine of a sequence of commands: *sit! stay!* (long torturous pause), *okay!* In that routine, one might have to wait an awfully long time before being given the okay to go at the food. Dogs put up with this game of ours with admirable self-possession. But if the owner starts chatting with someone else in the room—busying himself with someone else's attention—why, the game is off.

Lest you think that you can use this knowledge to trick your dog into behaving himself while you are at work by simply pretending to be home with him—over speakerphone or video—one experiment brings very disappointing news. When a life-sized video image (in visible digital) of the owner was displayed before dogs, they disobeyed at levels befitting being home alone with no supervision. While they could use their video-owners' pointing hints to help find food, they didn't bother to follow many of their verbal commands. Dogs are dutiful, but more selectively dutiful when the owner is reduced to a videotape. You cannot hope to reduce your dog's lonely wailing by telling him to stop over the answering machine—but you might be able to tell him where he can find that treat you left out for him.

When you next visit the zoo, check in on the monkey cages. Maybe there are capuchin monkeys, quick-moving, tail-flaunting animals who leap easily and shriek piercingly. Or colobus, slow-moving leaf-eaters whose black-and-white coats often hide a small colobus clinging inside. Watch the male snow monkeys as they follow around the red-bottomed females. There is much to recognize here in our distant evolutionary cousins. We see their interests, their fears, their lusts. And most will notice and respond to you—most likely by moving farther away, or turning their heads to avoid your gaze. What is surprising is that dogs, so much less humanlike than these primates, are so much better at realizing what is behind our gaze, how to use it to get information or to their advantage. Dogs can see us as our primate cousins cannot.

Canine Anthropologists

I am I because my little dog knows me.
— Gertrude Stein

The dog's gaze is an examination, a regard: a gaze at another animate creature. He sees us, which might imply that he thinks about us—and we like to be considered. Naturally we wonder, in that moment of shared gaze, Is the dog thinking about us the way we are thinking about the dog? What does he know about us?

We are known by our dogs—probably far better than we know them. They are the consummate eavesdroppers and peeping Toms: let into the privacy of our rooms, they quietly spy on our every move. They know about our comings and goings. They come to know our habits: how long we spend in the bathroom, how long we spend in front of the television. They know whom we sleep with; what we eat; what we eat too much of; whom we sleep too much with. They watch us like no other animal watches us. We share our homes with uncounted numbers of mice, millipedes, and mites: none bothers to look our way. We open our door and see pigeons, squirrels, and assorted flying bugs;

they barely notice us. Dogs, by contrast, watch us from across the room, from the window, and out of the corner of their eyes. Their watching is enabled by a subtle but powerful ability that begins with simple vision. Sight is used to pay visual attention, and visual attention is used to see what *we* attend to. In some ways this is similar to us, but in other ways it surpasses human capacity.

The blind and the deaf sometimes keep dogs to see or hear the world for them. For some disabled persons, a dog may enable movement through a world they cannot navigate alone. Just as for the physically impaired dogs can act as eyes, ears, and feet, so also do they act as readers of human behavior for some autistic individuals. Persons with any kind of autism spectrum disorder are united by their shared inability to understand the expressions, emotions, and perspectives of other people. As the neurologist Oliver Sacks describes, for an autistic person who keeps dogs, the dogs may seem to be human-mind-readers. While an autistic person cannot parse a brow furrowed with concern, or interpret the rising tone indicating someone's fright or worry, the dog is sensitive to the mind-set behind them.

Dogs are anthropologists among us. They are students of behavior, observing us in the way that the science of anthropology teaches its practitioners to look at humans. As adults, we walk among other humans largely without examining them closely, socially trained to keep to ourselves. Even with those we know best, we might stop attending to the minute changes in their expressions, their moods, their outlooks. The Swiss psychologist Jean Piaget suggested that as children we are little scientists, forming theories about the world and testing them by acting. If so, we are scientists who hone our skills only to later neglect them. We mature by learning how people behave, but eventually we pay less attention to how others are behaving at every instant. We outgrow the habit of looking. A curious child stares with fascination at the stranger limping down the street: he will be taught this is not polite. A child might be enraptured by a swirl of fallen leaves on the pavement; by adulthood, he will overlook it. The child wonders at our crying, monitors our smiles, looks where we look; with age we are all still able to do all this, but we fall out of the habit.

Dogs don't stop looking—at the gimpy walk, at a rush of leaves tumbling down the sidewalk, at our faces. The urban dog may be bereft of natural sights, but he is rich in the odd: the drunken man swerving through a crowd, the shouting side-walk preacher, the lame and destitute. All get long stares from the dogs who pass them. What makes dogs good anthropologists is that they are so attuned to humans: they notice what is typical, and what is different. And, just as crucially, they don't become inured to us, as we do—nor do they grow up to be us.

DOGS' PSYCHIC POWERS DECONSTRUCTED

This attunement to us feels magical. Dogs are able to anticipate us—and, it seems, to know something essential about us and others. Is this clairvoyance? A sixth sense?

I am reminded of the story of a horse. At the turn of the twentieth century, the actions of the horse Hans, whose ironic sobriquet, "Clever Hans," has come to stand both for what he could not do, and as a warning against overattributions to animals, helped shape the course of animal cognition research for the next hundred years.

Hans, his owner claimed, could count. Shown an arithmetic problem written on a blackboard, Hans tapped out the sum on the earth with his hoof. Though he had been encouraged and reinforced, using straightforward conditioning, for tapping, this was not a rote response to predetermined questions: he was excellent at all sums, with novel problems, and even when the questioner was someone other than his trainer.

Such was the tenor of the time that this discovered, presumed latent cognitive ability of horses created a small furor. Animal trainers and academics alike were stumped as to how Hans was doing it. It almost looked as though there were no other explanation than that he was actually doing arithmetic.

Finally, the trick—an inadvertent trick, unknown even to his owner—was discovered by a psychologist named Oskar Pfungst. When the questioner himself was prevented from knowing the answer to the problem, Hans's math was wildly off. Hans was not counting, and he was not psychic; he was simply reading the behavior of his questioners. They unconsciously tipped him off to the answer through small bodily movements: leaning forward or away from the horse when he'd tapped the correct sum; relaxing their shoulders and muscles of their face; inclining ever so slightly until the answer was reached.

Clever Hans stood, and still stands, as a cautionary tale against assigning to animals abilities that could be explained by simpler mechanisms. But thinking about the dog's use of attention reminds me of Hans's skill. While Hans was not clever in the way advertised, he was remarkably clever at reading the inadvertent signals given by the persons quizzing him. Before an audience of hundreds, only Hans noticed his trainer leaning forward, his body tensing and relaxing, which Hans had figured out meant that he was to stop tapping his hoof. He attended to the very cues that had information: an attention far greater than the human spectators brought to the event.

Hans's preternatural sensitivity may have ensued, paradoxically, from other deficits. Since he presumably didn't have any notion about numbers or arithmetic, he was not distracted by those stimuli. By contrast, our attention to the seemingly salient details would lead us to miss the one clear indication of the answer.

An experimental psychologist I've met who does research with pigeons demonstrates this phenomenon when teaching his undergraduate class. He shows the students a series of slides of bar graphs with blue bars of various lengths set against a white background. The slides fall into one of two categories, he says: those that have some unspecified "x-ness" feature, and those that don't. He points out which slides are members of the x-ness category. He then puts it to the students to use the example slides to figure out what the conditions for x-ness are.

After many minutes of frustrated and failing attempts by the students, he reveals that pigeons trained on a set of members with x-ness can without fail tell of a new graph whether it satisfies this elusive criterion or not. The students shift uncomfortably in their seats. Still, not a person comes up with the answer. Finally, their professor fills them in: those slides that are mostly blue are members of the x-ness category; those that are more white are not.

The students are outraged: they've been outsmarted by pigeons. In running this test with my psychology classes, I find they also decry the task. Though no student has ever come up with the answer, they all later complain that the answer is unfair. They were looking for some complicated relationship between the bars—one consistent with the kinds of relationships between features that bar graphs are meant to represent. But there is none. "X-ness" is simply "more blue." Only pigeons, blissfully unaware of bar graphs, saw them for their colors and perceived the true categories.

What dogs do is a version of what Hans and the pigeons did. Anecdotal tales of this kind of phenomenon abound. One trainer of search-and-rescue dogs put his hands on his hips in exasperation when the dog was following the wrong path. Another rubbed his chin uneasily. In both cases, the dogs learned to use the cues of their trainers as information that they were off the trail. (The trainers had to be trained to tone down their cuing.) As we look for the more complex explanation for an event, or for others' behavior, we may overlook clues that dogs see naturally. It is less extrasensory perception than the well-added sum of their ordinary senses. Dogs use their sensory skills in combination with their attention to us. Without their interest in our attention, they would not perceive the subtle differences in our strides and body postures and stress levels as important bits of information. It allows them to predict us and to reveal us.

READING US

The dog observes us, thinks about us, *knows* us. Do they then have some special knowledge about us, born of their attention to us and to our attention? They do.

In a non-verbal way, dogs know who we are, they know what we do, and they know some things about us unknown to ourselves. We're knowable by look, and even more so by our smells. Over and above that, how we act defines who we are. Part of my recognition of Pump is not just of her visage; it is of her walk: her slightly off-kilter, jaunty trot with her droopy ears bouncing in step. For dogs, too, the identity of a person is not just how she smells and looks; it is how she moves. We are recognizable by our behavior.

Even our most ordinary behavior—walking across a room in our characteristic style—is chock full of information that the dog can mine. All dog owners watch their pups' growing sensitivity to the rituals that precede going for what in many dog-peopled households is called a W-A-L-K.* Dogs quickly learn to recognize shoe donning, of course; we come to expect that grabbing a leash or a jacket will clue them in; a regular walk time explains their prescience; but what if all you did was look up from your work or rise from your seat before your dog was on to you?

If done suddenly, or if you cross the room with a purposeful stride, an attentive dog has all the information he needs. Habitual watcher of your behavior, he sees your intent even when you think you are giving nothing away. As we've seen, dogs are very sensitive to gaze and thus to changes in our gaze. The difference between a head lifted up or angled down, away from them or toward them, is large for an animal so sensitive to eye contact. Even small movements of the hands or adjustments of the body attract notice. Spend three hours looking at a computer screen, hands tethered to the keyboard, then look up and stretch your arms overhead—this is a metamorphosis! The redirection of your attention is clear—and a hopeful dog can easily interpret it as a prelude to a walk. An acute human observer

*Spelling the word instead of saying it is, of course, usually futile. Dogs can also learn the connection between the cadence of a spelled word and a subsequent walk, even if the latter does not immediately follow the former. On the other hand, used in an unlikely context—say, sitting in the bath—the spelled word will not evoke much interest. Chances are slim you're about to up and take a walk when naked and sudsy.

would notice this, too, but we rarely let others oversee us so closely in our daily affairs. (Nor do we find it terribly interesting to so oversee.)

Their facility at anticipation of our actions is part anatomy and part psychology. Their anatomy—all those rod photoreceptors—allows them a millisecond head start on noticing motion. They react before we see that there is something to react to. The critical psychologies are of anticipation—predicting the future from the past—and of association. Familiarity with your typical movements is necessary in order to so anticipate you: a new puppy might not be tricked by a feigned tennis-ball toss, but with age he will be. Even without familiarity, dogs are skilled at making associations between events—between the arrival of one's mother and the delivery of food; between a shift in your focus and the promise of a walk.

Dogs pick up the theme of our quotidian habits, and thus are especially sensitive to variations in them. Just as we often take the same route to our cars, to work, to the subway, we take our dogs on similar walks. Over time, they learn the route themselves, and can anticipate that we turn left past the hedgerow and make a sharp right at the corner with the fireplug. If we introduce a new detour on our way home, even an unnecessary one—circling around the block an additional go—dogs adjust to the new route after just a few outings. And they even begin heading in the direction of the detour before their owner makes any movements in that direction. This makes them fine, cooperative walking companions—better than many humans I perambulate the city with, whom I constantly knock into as I lead them on a preferred route.

The complement of dogs' anticipatory prowess is their purported character-reading ability. Plenty of people let their dogs choose their potential romantic partners. Others declare their dog a good judge of character, able to spot a duplicitous person, a bad sort, on first meeting. They may seem to recognize someone who is not to be trusted.* What this ability might come down to is their close looking at our looking. If you feel hesitant about an approaching stranger, you reveal it, however unintentionally. Dogs are, as we have seen, sensitive to the olfactory changes that come with stress; they can also notice tensing muscles and the auditory change of quick breathing or gasping. (These physiological changes are among those measured by lie detectors: one might imagine that a trained dog could substitute both for the machine and its technician.) But they will let their visual ability trump these notes when assessing a new person or trying to solve a problem. We all have characteristic behaviors we display when angry, nervous, or excited. "Untrustworthy" people often glance furtively in conversation. Dogs notice this gaze. An aggressive stranger may make bold eye contact, move unnaturally slowly or quickly, or veer oddly from a straight path before doing any actual aggression. Dogs notice the behavior; they react viscerally to the meeting of eyes.

One winter we took a trip north, to a place of assertive winter and genuine cold, and were treated to a large snowstorm. We pulled out sleds, found a great big hill, and proceeded to plow an erratic track down it. Pump was suddenly overcome, and ferociously pursued us on each ride downhill, biting, grabbing, and

*While dogs may in fact distinguish between people behaving subtly differently, one suspects that anyone using their dogs this way might be susceptible to what psychologists call *confirmation bias:* noticing just that part of their dog's response that supports their own theories about the person. Does the gentleman seem a bit untrustworthy to you? And yes, look how your dog growled at him once: that settles it. Dogs become amplifiers of our own beliefs; we can attribute to them that which we think ourselves.

growling at our faces. When it was my fast-moving, snow-covered face being attacked, I couldn't stop her for all my laughing. She was playing, but it was a play I have not seen before: tinged with real aggression. When I managed to rise and shake off the covering of snow I'd gathered on the way down, she calmed at once.

Does this clairvoyance mean that dogs can't be fooled? No. They are astute watchers, but they are not mind readers, nor are they immune from being misled. I was *changed* for Pump when I leapt on that sled: I was horizontal; I was enrobed in snow gear and snow; and most critically, I was moving entirely differently. I was suddenly a smoothly moving, high-speed prey animal, not an upright, ambling companion.

My dog may have a particular interest in sledders, but her behavior is similar to many other dogs' chasing behaviors. Dogs often chase bicyclists, skateboarders, RollerBladers, cars, or runners. The general-purpose answer given for why they do this is usually that they have an instinct to chase prey. This answer is not entirely wrong, but it is mightily incomplete. It is not quite that dogs think of these objects or persons as "prey," per se. Your motion reveals another dimension to you: *you roll! quickly!* It is an attribute that alters you in the dog's eyes, which are especially responsive to a certain kind of motion. Mounting and riding a bicycle, you have not turned to prey—as indicated by the fact that your dog greets you, not eats you, on dismount. Their responsive sensitivity probably evolved as a prey-detection tactic, but it will be applied variously. It lends to the dog's experience an additional way to interpret objects and animals in the environment. That way is by the quality of their motion.

There are shared components of sledding, bicycling, or running: a person is moving in a certain way—smoothly and quickly. Walkers are moving, but not quickly: they are not chased. Pump did not recognize me sledding because ordinarily, much as I would like to think otherwise, I am not particularly smooth nor quick in my motions. There is an excess of vertical movement in my walk; I weave to and fro; I gesture a lot—all frivolous in making forward progress.

To stop a dog pursuing a bicycle with a predatory glint in his eye, one can simply interrupt the illusion: stop the bike. The chasing impulse triggered by the visual cells that detected the motion will itself let up. (The hormones involved in the arousal of barking and chasing such a smooth and quick mover may still be coursing through his system, though, for a few minutes.)

Science has confirmed the importance of behavior in identity. Our identities, who we are, are defined partly by our actions, so we can examine how actions inform recognition of personal identity. In one experiment, dogs showed that they have no difficulty distinguishing friendly and unfriendly strangers: those demonstrating different identities. To do this, the experimenters divided participants into two groups and asked members of each group to behave in a prescribed manner. Friendly behavior included walking at a normal speed, talking to the dog in a cheerful voice, and gently petting the dog. Unfriendly behavior included actions that could be interpreted as threatening: an erratic, hesitating approach combined with staring at the eyes of the dog without talking.

The main result of the experiment is not all that surprising: the dogs approached the friendly and avoided the unfriendly. But there's a hidden gem in the experiment. The key trial was this: How did the dogs act when a formerly friendly person suddenly acted threatening? The dogs acted variably: For some, the person was now a different kind of person altogether—an unfriendly one, her identity changed. To others, the olfactory recognition of the stranger who had been friendly trumped this new odd behavior.

These people began as strangers to the dog, but over the course of the sessions, dogs became familiar with the various people: they became "less strange." Their identity was defined in part by their smell and in part by their behavior.

ALL ABOUT YOU

The combination of dogs' attention to us and their sensory prowess is explosive. We have seen their detection of our health, our truthfulness, even our relation to one another. And they know things about us at this very moment that we might not even be able to articulate.

The results of one study indicate that dogs pick up on our hormonal levels in interaction with them. Looking at owners and dogs participating in agility trials, the researchers found a correlation between two hormones: the men's testosterone levels, and the dogs' cortisol levels. Cortisol is a stress hormone—useful for mobilizing your response to, say, flee from that ravenous lion— but also produced in conditions that are more psychologically than mortally urgent. Increases in the level of the hormone testosterone accompany many potent elements of behavior: sex drive, aggression, dominance displays. The higher the men's pre-agility-competition hormonal levels, the higher the increase in the level of stress in the dogs (if the team lost). In a sense, the dogs somehow knew that their owner's hormonal level was high, by observing behavior or through scent or both—and they "caught" the emotion themselves. In another study, dogs' cortisol levels revealed that they were even sensitive to the *style* of play of human playmates. Those dogs playing with people who used commands during play—telling the dog to sit, lie down, or listen—wound up with higher postplay cortisol levels; those playing with people who played more freely and with enthusiasm had lower cortisol at play's end. Dogs know, and are infected by, our intent, even in play.

Being known and predicted by our dogs is no small part of our fondness for them. If you have experienced an infant's first smile at you as you approach, you know the thrill of being recognized. Dogs are anthropologists because they study and learn about us. They observe a meaningful part of our interaction with each other—our attention, our focus, our gaze; the result is not that they can read our minds but that they recognize us and anticipate us. It makes the infant human; it makes the dog vaguely human, too.

Noble Mind

It's dawn and I try to sneak out of the room without waking Pump. I can't see her eyes, so dark they're camouflaged against her black fur. Her head rests peacefully between her legs. At the door I think I've made it—tiptoed and breath-held to avoid her radar. But then I see it: the swell of her lifted eyebrows tracking my path. She's on to me.

The dog, as we've seen, is a master looker, a skilled user of attention. Is there a thinking, plotting, reflective mind behind that look? The development of the human infant's looking into using attention marks the blossoming of the mature human mind. What does the dog's looking tell us about the dog mind? Do they think about other dogs, about themselves, about you? And the time-worn but still unanswered question of dog minds: Are they smart?

DOG SMARTS

Dog owners, like new parents, always seem to have a handful of stories at the ready describing how smart their charges are.

Dogs, it is claimed, know when their owners are going out, and when they are coming home; they know how to hoodwink us and they know how to beguile us. News reports buzz with the latest discovery of the intelligence of dogs: of their ability to use words, count, or call 911 in an emergency.

To verify this anecdotal impression, some have designed so-called *intelligence tests* for dogs. We're all familiar with intelligence tests for humans: pen-and-paper creations that require you to solve SAT-like problems of word choice, spatial relationships, and reasoning. There are questions that test your memory, your vocabulary, your declining math skills, and your simple pattern-finding ability and attention to detail. Even putting aside whether the result is a fair assessment of intelligence, the design does not translate obviously to testing dogs. So revisions are made. Instead of tests of advanced vocabulary, there are tests of simple command recognition. Instead of repeating a list of digits read aloud, a dog may be asked to remember where a treat was hidden. Willingness to learn a new trick may replace the ability to figure complex sums. Questions loosely mimic experimental psychology paradigms: of object permanence (if a cup is placed over a treat, is it still there?), learning (does your dog realize what foolish trick you desire him to do?), and problem solving (how can he get his mouth on that food you've got?).

Formal studies of groups of dogs on these kinds of abilities—mostly cognition about physical objects and the environment—yield what at first seem to be unsurprising results. By bringing dogs to a field baited with treats and timing dogs' speed in finding them, researchers have confirmed that dogs use landmarks to navigate and find shortcuts. This behavior is consistent with what their wolflike ancestors would probably have done in finding food and finding their way. Dogs are, of course, pretty good at all tasks that involve getting themselves to food. Given a choice of two piles of food, dogs have no trouble choosing the larger one—especially as the contrast between them grows. Turn a cup over a bit of food and dogs go right for it, knocking the cup and revealing the treat. Dog subjects have even learned how to use a simple tool—pulling a string—to get an attached biscuit that was otherwise out of reach.

But dogs don't pass all the tests. They typically make lots of mistakes when presented with piles of three versus four biscuits, or of five and seven: they choose the smaller amounts just as often as the larger. And they develop preferences for piles on the left or the right, which lead them to make even more blatant errors. Similarly, their skill at finding hidden food gets worse as the hiding gets more complicated. And their tool use also starts to look less impressive as the trials get trickier. When there are two strings, and only the more distant one is attached to an alluring biscuit, dogs nonetheless go for the nearer string, the one attached to nothing. They don't seem to understand the string as a tool: as a means to an end. Indeed, they may have succeeded in the original case simply by pawing and mouthing at the problem until accidentally solving it.

A dog owner tallying her dog's score in these dog intelligence tests might find that he's scoring closer to *Dim but happy* than *Top of the obedience class.* Is that it, then? Is he not smart after all?

A closer look at the intelligence tests and the psychological experiments reveals a flaw: they are unintentionally rigged against dogs. The flaw is in the experimental method, not in the experimented dog. It has to do with the very presence of people—experimenters or owners. Let's look more closely at a typical experimental setup. It might begin as follows: A dog is sitting at attention and being restrained by a leash. An experimenter comes before him and shows him a great new toy. This dog loves new toys.* The toy and a bucket are clearly shown to the dog, the toy is put into the bucket, and then the experimenter disappears with the booty behind one of two screens in the room. She returns with the bucket—emptied of its treat. This turns out not to be a cruel hoax, but a standard test of *invisible displacement:* wherein an object is *displaced*—moved to another location—*invisibly*—out of sight. This test has been regularly run with young children since Piaget proposed it as representing one of the conceptual leaps that infants make on their way to becoming incorrigible teenagers and then adults capable of having infants of their own. In this case, the conceptual understandings are of the continued existence of objects when they are out of sight—called *object permanence*—and some notion of that object's trajectory and continued existence in the world. If someone disappears behind a door, we realize not only that they still exist when we can't see them, but that we might find them by looking behind that door. Children master object permanence before their first birthday, invisible displacement by their second. Since Piaget reified this representational understanding as a stage in infant cognitive development, it is a standard test that is run with other animals, to see how they compare to little people. Hamsters, dolphins, cats, chimpanzees (who reliably pass), and chickens have all been tested. And dogs.

Dogs' performance is mixed. Oh, sure, if the test is run simply as described, then they have no trouble looking behind the screen for the toy. It looks as though they've passed the test. But

---

*Dogs have a preference for novel objects—*neophilia*. One study found that when asked to retrieve an unspecified toy from a pile of familiar and new toys, dogs spontaneously chose the new toys over three-quarters of the time. This penchant for the new might explain why when two dogs carrying sticks meet in a park, they often simultaneously drop the stick they've been proudly toting around in order to try to grab the pride of the approaching dog.

complicate the scenario a little—carry the container behind two different screens, taking the toy out after the first screen *and showing them that you have done so* before going behind the second screen—and dogs fail: they race to the second screen first, where the toy clearly is not. Other test variations also result in dogs suddenly looking less smart in their searching. We could conclude that here too the dogs appear to be less than genius. Once the toy is out of sight, it may quickly fall out of mind.

But the very fact that dogs do succeed, sometimes, renders that conclusion suspect. Instead their behavior points to two explanations. First, it is likely that dogs remember the toy, but do not engage in detailed consideration of what its path might be when it vanishes. Though some dogs are indisputably keen to keep track of a toy, dogs nonetheless regard objects in their environment very differently than humans do. Significantly, what wolves and dogs do with objects is limited: some objects are eaten, and some are played with. Neither interaction requires complex rumination on the object. Dogs realize when a previously treasured object is missing, but needn't mull over possible stories for what happened to it. Instead they just start looking for it, or wait for it to show up.

The second explanation is more far-reaching. It appears that the very skill at social cognition that is their triumph as a companion to humans contributes to the dogs' failure at this and other physical-cognition tasks. Show your dog a ball, then conceal it from him while you place it under one of two overturned cups. Faced with the cups, and assuming he can't smell it out, a dog will look under either cup at random: a reasonable approach when he has nothing to go on. Lift one cup to reveal a peek of the ball underneath, and you won't be surprised that when allowed to search, your dog will have no trouble looking under that cup. But give a peek under the cup holding nothing, and researchers found that dogs suddenly lose their logic. They search first under the empty cup.

These dogs were stymied by their own skill. When presented with a problem of any kind, dogs cleverly look to us. Our activities are sources of information. Dogs come to believe that our actions are relevant—often leading, we might note, to some interesting reward or even food. So if an experimenter ducks behind a second screen, as she does in the more complicated invisible displacement tasks, why, there might be something of interest behind that screen. If she lifts up an empty cup, that cup becomes more interesting simply because of her attention to it.

If the social cues are diminished in the tests, dogs perform much better. When experimenters handle both cups even when showing the dog the empty one, dogs regain their heads. They see the empty cup, and by deduction search under the other cup, which holds the hidden ball. Similarly, dogs who are less well socialized—such as *yard dogs* kept outside for most of their hours—also set right to the problem, while dogs who live inside the house more often plead quietly with their owners to help.

If we revisit some of the problem-solving tests on which wolves performed so much better than dogs, we now see that the dogs' poor performance can there too be explained by their inclination to look to humans. Tested on their ability to, say, get a bit of food in a well-closed container, wolves keep trying and trying, and if the test is not rigged they eventually succeed through trial and error. Dogs, by contrast, tend to go at the container only until it appears that it won't easily be opened. Then they look at any person in the room and begin a variety of attention-getting and solicitation behaviors until the person relents and helps them get into the box.

By standard intelligence tests, the dogs have failed at the puzzle. I believe, by contrast, that they have succeeded magnificently. They have applied a novel tool to the task. We are that tool. Dogs have learned this—and they see us as fine general-purpose tools, too: useful for protection, acquiring food, providing companionship. We solve the puzzles of closed doors and empty water dishes. In the folk psychology of dogs, we humans are brilliant enough to extract hopelessly tangled leashes from around trees; we can magically transport them to higher or lower heights as needed; we can conjure up an endless bounty of foodstuffs and things to chew. How savvy we are in dogs' eyes! It's a clever strategy to turn to us after all. The question of the cognitive abilities of dogs is thereby transformed: dogs are terrific at using humans to solve problems, but not as good at solving problems when we're not around.

LEARNING FROM OTHERS

Yesterday Pump learned, courtesy of a pet supermarket's automatic doors, that when you walk toward walls, they open and let you pass through them. Today she unlearned it, in a spectacularly poignant display.

Once a problem is solved—a hidden treat is unearthed, an unjustly closed door is opened—with or without a person's help, the dog is quickly able to apply that same means to solve it again and again. He has identified a state of affairs, fashioned a response, and realized the connection between that problem and that solution. This is both his triumph and, at times, our misfortune. One success at jumping right onto the kitchen counter to get to the origin of that pleasing cheese odor will be followed by much jumping-on-counters. If you provide a sitting dog with a biscuit for sitting politely, expect to be inundated by polite sits. With this in mind it is easy to understand the admonishment that in training a dog you must reward only those behaviors you desire the dog to repeat endlessly.

Such is the dog's mastery of what in psychological circles is termed *learning*. There is no doubt that dogs can learn. It is the natural workings of any nervous system to adjust its actions over time in response to experience—and of every animal with a nervous system to thereby learn. Under the heading "learning" comes everything from the associative learning used in animal training, to memorization of a Shakespearean monologue, to finally understanding quantum mechanics.

Dogs' easy mastery of new procedures and concepts presumably stops prior to grasping what a quark is. What they learn is neither academic nor scholastic. Still, most of what we ask that dogs learn can only be described as capricious and arbitrary. Surely any animal recently wild will learn how to get its mouth on food. But typically the things we want dogs to learn—to obey—bear little connection to food. We ask dogs to change posture (to sit, jump up, stand up, lie down, roll over), to act in a very specific way on an object (get my shoes, get off the bed), to start or stop a current action (wait, no, okay), to change mood (cool it, go get him!), to move toward us or move away from us (come, go away, stay). This may not be quantum mechanics, but it is just as bizarre to these distant moose hunters. Nothing in a wild animal's life prepares him to be asked to maintain the state of holding his rump on the ground, unmoving, until released by your cheery *okay!* It is notable that dogs can learn these seemingly arbitrary things at all.

PUPPY SEE, PUPPY DO

One morning, on awakening lying on my belly, I pulled my arms over my head, stretched my legs into pointed toes, and pulled myself up onto my forearms. Aside me Pump stirred, and matched me move for move: she tensed her front legs, stretched them well out in front of her, then straightened her back legs, too, pulling herself forward into uprightness. Now we greet each other every morning with parallel wakening stretches. Only one of us swings her tail.

Even more interesting than learning commands would be the ability to learn by merely watching others—other dogs or even people. We know dogs can learn from our instruction, but can dogs learn from our example? It would seem to behoove a social animal like the dog to look to others for information about how best to negotiate the world. In many cases, though, the answer to this question is clearly *no:* dogs have plenty of opportunity to see us eating politely at the table—yet they never spontaneously pick up knife and fork and join us. Overhearing us talking is insufficient to get them talking; their only interest in clothes seems to be chewing them, not donning them. Amply exposed to our activities, dogs don't seem to know how to imitate us.

This is not a failing, though it would distinguish them from members of our own species, consummate imitators that we are. As children and into adulthood, we goggle at each other to see what to wear, what to do, how to act, and how to react. Our culture is built on our keenness in observing others act to learn how to behave ourselves. I need only see you opening a tin can with a can opener once before I can do it myself (one hopes). The stakes are higher than they might first seem, for success at imitation not only gets you the contents of the opened can, it is an indication of a complex cognitive ability. True imitation requires that you not merely can see what another is doing, not simply that you see how the means lead to an end, but also that you translate others' actions into your own actions.

In that case, dogs are not true imitators, for even after thousands of demonstrations with the can opener, no dog has shown an interest: the opener's functional tone is mute for them. But this is not a fair comparison, you might complain: dogs simply haven't the thumbs, nor the dexterity they allow, to operate can openers or cutlery. Similarly, they haven't the larynx for speech nor the need for clothing. And your complaint would be fair: the question is really if the dogs can be taught, by demonstration, how to do something new—not whether they are mini-humans.

Watch dogs interact for ten minutes and you will see what looks like imitation: one dog flaunts a gloriously large stick; the other finds a stick of his own and flaunts it back. If one dog finds a spot for digging, others will soon join him at the growing hole; one dog's discovery that he can swim leads another dog to self-baptize, suddenly finding himself swimming, too. By watching others, dogs learn the special pleasures of mud puddles and of bushwhacking through brush. Pump uttered nary a peep until one of her regular dog companions began barking at squirrels. All at once, Pump too was a squirrel-barker.

The question, then, is whether these are cases of true imitation, or of something else. The something else that it might be is opaquely called *stimulus enhancement*. A minor incident involving birds and home-delivered milk in mid-twentieth-century Britain demonstrates this phenomenon best. At the time, doorstep milk delivery was commonplace in Britain, and homogenization was not. Thus dawn found foil-capped bottles of separated milk idling unattended on front porches, the cream nearest to the top of the bottles. Up at dawn with the delivery men is much of Britain's bird population, for dawn is a propitious time to sing. One bird, the small blue tit, made a discovery: the foil on the bottles was susceptible to being pecked through, revealing a rich creamy drink just below. A few reports of vandalized milk bottles were lodged, soon a spate more, then a plague of them. Hundreds of birds had learned the milk-bottle trick. Cross with their skimmed milk, the Britons were not long in finding the culprits. For us, the question is not who but how: How did this discovery spread among the blue tits? Given the rapidity with which it spread, it seemed likely that some birds observed others getting the cream, and imitated them doing so. Clever, pudgy little birds.

By providing a captive population of chickadees with a similar setup, one group of experimenters observed the phenomenon recur step-by-step. Their studies suggest a more likely explanation than imitation. Instead of carefully observing and assimilating all that the first, cream-pilfering bird was doing, other birds simply saw that he was atop the bottle. This may have attracted them to the bottles. Once landed on the bottle tops, by doing a natural behavior—pecking—they discovered the foil's puncturability themselves. In other words, they were drawn to a *stimulus,* the bottle, by the first bird's presence. Its presence enhanced the likelihood that they too would become cream stealers, but it did not demonstrate how to do so.

This may seem nitpicky, but there is an important difference at work here. In a case of stimulus enhancement, I see that you are acting in some unspecified way on the door, after which it opens. If I amble over to the door and kick it, hit it, and otherwise maul it, I might get it to open, too. In a case of imitation, I watch exactly what you are doing with the door and reproduce just those actions—the seizing and turning of the knob, the application of pressure after turning, and so on—that lead to the desired outcome. I can do that because I can imagine that what you are doing is somehow related to your goal, your desideratum: to leave the room through the door. The blue tits, on the other hand, need not have been thinking about what the milk bottle tits wanted—and probably were not.

MORE HUMAN THAN BIRD

Dog researchers wanted to test whether the stick-flaunting dogs are acting more like a blue tit or more like a human being. The first experiment was designed to determine if dogs would imitate humans in a situation in which the people were acting to attain some desired object. The researchers were asking, in essence, whether dogs can understand that a person's actions function as a demonstration that can be followed if the dog is otherwise unsure how to get that desired object himself.

They set up a simple experiment in which a toy or a bit of food was placed in the crook of a V-shaped fence. The dog was seated on the outside of the point of the V, and was given a chance to try to retrieve the food. He couldn't go straight through or over the fence, but both routes around the fence— around the left stem or the right stem—were equally long, so equally good. When given no demonstration of how to get around the fence, the dogs chose randomly, preferring neither side, and eventually making their way to the inside of the V. But when given a chance to watch a person walking around the *left* side of the fence toward the reward—a person actively talking to the dog along the way—the observing dogs changed their behavior outright: they also chose the left side.

It looks as though these dogs were imitating. And what they learned by imitating stuck: when a shortcut through the fence was later introduced, they maintained the route they had learned by watching, ignoring the shortcut. The researchers ran a handful of other trials to be clear what exactly it was that the dogs were doing. They were not simply navigating by smell: laying down a scent trail on the left arm of the fence did not induce dogs to follow it.* Instead it had something to do with understanding others' actions. Simply watching someone quietly walk around the fence was insufficient to get the dogs to follow the person's route: the person had to be calling the dog's name, grabbing attention, yammering away. Watching another *dog* who had been trained to retrieve the reward by the left-hand route also prompted observing dogs to go left.

This result showed that dogs can see others' behavior as a demonstration of how to get to a goal. But we know from experience with our dogs that not every relevant behavior we do is seen as a "demonstration." Pump may watch me navigate around strewn chairs, books, and clothes piles as I head to the kitchen, but she will herself charge right through piles to take the quickest route. Other tests are necessary to determine if dogs are truly putting themselves in our shoes, and not just prone to *follow that human,* wherever we go.

---

*The dogs' ambivalence about the scent trail may at first seem surprising, given all our talk about their olfactory skills. But simply being able to smell a trail doesn't mean that they *use* this ability all the time. Often dogs need to be trained to be attentive to particular scents.

Two experiments have tested just this imitative understanding. The first asked what exactly dogs see in others' behavior: the means or the end. A good imitator would see both, but would also see if the particular means isn't the most expedient way to the end. From a young age, human infants can do just that. They will religiously imitate—sometimes to a fault*—but they can also be astute. For instance, in one classic experiment, after watching an adult turn on a light in an unusual way—with his head—the infant subjects could imitate this novel action, if asked to do so. But they did not spontaneously imitate if the adult was grasping something in his hands, making him unable to use them to turn on the light: the infants used their hands, reasonably enough. If the adult held nothing in his hands, infants were more likely to turn on the light with their heads, too—inferring, perhaps, that there must be good reason, besides one's hands being full, for this new maneuver. They seemed to realize that the adult's actions *could* be imitated, and they selectively imitated them only insofar as it seemed necessary to do so.

In the dog variation on this paradigm, a wooden rod taking the place of a light, one "demonstrator" dog was taught to press the rod with his paw to release a treat from a spring-loaded dispenser. The researchers then had the demonstrator dog perform his newfound trick in front of other dogs who were being

---

*My favorite example of the child's overimitation comes from an experiment that psychologist Andrew Whiten and his colleagues ran using a locked box with a tempting piece of candy inside. They were curious if three- to five-year-old children could imitate the particular means experimenters demonstrated to unlock the box (involving twisting out rods fit through barrel openings). The children watched, captivated, and were then handed the re-locked box. Whiten found that the children almost all imitated—and the youngest children *over*imitated—twisting the rod not two or three but sometimes hundreds of times before pulling it out. What they did not yet understand was exactly what part of the (twisting) means was necessary to get the (candy-yielding) end.

restrained to watch. In one trial the demonstrator pressed the rod while holding a ball in his mouth; in the other, he had no ball. Finally, the observer dogs were let at the apparatus.

It should be noted that dogs are not naturally drawn to mechanical dispensers, even ones with wooden rods. And *pressing* is not the first approach of most dogs when facing a problem: dogs can use their paws handily, but they typically go at the world mouth first and paws second. Though they can be trained to push or press an object, dogs' first approach at an object such as this one is not one of intuitive understanding. They will bump it, mouth it, knock into it. If they can, they will push it over, dig at it, jump on it. But they do not consider the scene for a moment and then calmly press the rod. Thus the first approach of the observer dogs was particularly interesting: Would the demonstration change their behavior?

These dog subjects behaved just like the human infants with the light switches: The group that saw the demonstration with no ball imitated faithfully, pressing the rod to release the treat. The group that saw the demonstrator acting while holding a ball in his mouth also learned how to get the treat, but used their (ball-less) mouths instead of paws.

That the dogs so imitated is remarkable. This is no mere mimicry, copying for copying's sake. Nor is it just an attraction to the source of activity. It looks more like the behavior of an animal who is considering what another animal is doing: what his intention is, and how—or how much—to reproduce that behavior themselves, if they have the same intent.

If these experiments represent the performance of all dogs, it looks as though we could say that dogs are, at the very least, able to learn by watching others in particular social contexts— when food is at stake, for instance. One final experiment suggests something even more impressive: that dogs may actually understand the *concept* of imitation. The single subject, an assistant dog trained to work with the blind, had already learned by operant conditioning to do a number of non-obvious actions on command: to lie down, turn around in a circle, put a bottle in a box. What the experimenters wondered was whether he could do these actions not just to a command, but after seeing someone else do the action themselves. Sure enough, the dog ably learned to turn around in a circle not after the *Turn around in a circle* command, but on simply seeing a human do such a thing, followed by the imitation request *Do it!* They then examined what he would do when seeing a human do a new, completely odd action, such as running off to push a swing, tossing a bottle, or suddenly walking around someone else and returning to their starting spot.

He did it. It was as though this dog had learned the concept *imitate,* and, given that notion, could apply it more or less in any direction. To do this, he had to map his body onto a human's: where a person tossed a bottle by hand, the dog used his mouth; he used his nose to push the swing. This is not the final word about imitation (just ask your dog to copy your swing-pushing, and you can see how results do not always generalize), but these dogs' abilities are suggestive of something besides mindless mimicry. Dogs may be enabled to imitate by the same ability— almost compulsion—to look at us that allows them to use us to learn how to act. That is what I see in Pump's morning stretch alongside me.

THEORY OF MIND

I open the door stealthily and Pump's there, not two feet away, walking toward the rug with something in her mouth. She stops in her tracks and looks over her shoulder at me, her ears down, her eyes wide. In her mouth is an unidentifiable curved form. As I slowly approach, she wags low, ducks her head, and in the moment that she opens her mouth to get a better grip on her find I see it: the cheese left out on the counter to warm. The brie. The entire enormous round of brie. She gulps two gulps and it's gone, down the gullet.

Think of the dog caught in the act of stealing food from the table . . . or looking at you squarely in the eyes with a plea to go out, be fed, be tickled. When I see Pump, mouth full of brie, seeing me, I know she's going to make a move; when she sees me seeing her, does she know I'm going to try to thwart it? My strong impression is that she does: the moment I open the door and she looks at me, we both know what the other is going to do.

The study of animal cognition reaches its pinnacle in addressing just this kind of scene: one raising the question of whether an animal conceives of others as independent creatures with their own, separate minds. This ability seems more than any other skill, habit, or behavior to capture what it is like to be a human: we think about what others are thinking. This is called having a *theory of mind*.

Even if you've never heard of theory of mind, chances are you nonetheless have a very advanced one. It allows you to realize that others have perspectives different from your own, and therefore have their own beliefs; different things they know and don't know; a distinct understanding of the world. Without one, others' behavior, even the simplest acts, would be utterly mysterious, arising from unknown motivations and leading to unpredictable consequences. Trying to guess what a man approaching you, mouth agape, arm raised high, hand waving frantically, is going to do is greatly aided by having a theory of mind. It's called a *theory* because minds are not directly observable, so we extrapolate backward from actions or utterances to the mind that prompted that act or remark.

We aren't born thinking about others' minds, of course. It is quite likely that we aren't born thinking about much at all, even our own minds. But each normal child develops a theory of mind eventually, and it appears that it is developed through the very processes discussed so far: through attending to others, and then noticing their attention. Children with autism often don't develop some or any of these precursory skills: they may not make eye contact, point, or engage in joint attention—and many don't seem to have a theory of mind. For most people, it is but one large theoretical step from an awareness of the role of gaze and attention to realizing that there is a mind there.

The gold standard experiment for theory of mind is called the *false belief* test. In this design, the subject, typically a child, is presented with a minidrama played out by puppets. One puppet places a marble in a basket in front of her, in full view of the subject and a second puppet. Then the first puppet leaves the room. Promptly, the second puppet wickedly moves the marble over to her basket. As the first puppet returns, the subject is quizzed: Where will the first puppet go looking for her marble?

By age four, children answer correctly, realizing that they and the puppet know different things. Before that age, though, children surprisingly and unambiguously fail. They say the puppet will look for the marble where the marble actually is—in the second basket—showing that they aren't thinking about what the first puppet really knows.

To design a verbal false belief task for animals, who cannot be expected to communicate their answers (nor be engaged by a puppet marble-switching drama) is nigh impossible, so nonverbal tests have been developed. Many take their cue from anecdotal reports of compellingly mindful animal behavior seen in the wild: of deception or clever competitive strategies. Chimpanzees are the most frequent subjects since, as close relatives to humans, one might expect that they would have the most similar cognitive abilities.

While the results with chimpanzees have been equivocal, lending credibility to the notion that only humans have a fully developed theory of mind, a wrench has been thrown in the experimental works. That wrench is the dog: whose attention to attention, whose seeming mind reading, looks anecdotally just like what we call acting with a theory of mind. To go from my living-room theorizing about a dog's understanding of mind to solid scientific standing, researchers have begun to run dogs on the same tests used with chimps.

THEORY OF DOG MIND

Here's what one dog, an unsuspecting experimental subject, found awaiting him at home one day. Instead of the usual ready availability of his favorite tennis balls, every ball in the house had been collected, and an extra lot of people were standing about gazing at him. Fine so far: Philip, the three-year-old Belgian Tervuren in question, didn't freak out. But he might have been bemused when, one by one, the balls were shown to him and then placed in one of three boxes and locked up. This was new stuff. Whether game or threat, what was clear was that the balls were being methodically placed somewhere other than his favorite place: right in his mouth.

When released by his owner, Philip went, naturally, straight toward the box where he saw a ball hidden, and he nuzzled the box. This turned out to be the right thing to do, for it prompted the humans to exclaim merrily, open the box, and give him the ball. Despite just having his mouth on the ball, the dog found that the people around him kept taking it away and securing it in one or the other box—so he kept playing along. Then they started locking the boxes and putting the key elsewhere, so the whole thing took even longer after he selected the right box: someone must find the key, bring it to the box, and open it. The final twist involved one person who locked the box, hid the key, then left the room. Another person came in—surely one who, like all other people around, would be able to use these key-things to open these lock-things.

This was the moment the experimenters were waiting for: they wanted to know if the dog saw the new person as unknowledgeable about the location of the key. If so, then not only should Philip indicate which box has the beloved ball, he should also help the person find the key that would enable access to that ball.

On repeated trials, that's more or less what the dog did: ever patient, Philip looked toward the spot where the key had been hidden, or headed that way. Note that he didn't actually take it in his mouth and open the box: that'd be some trick, but even the most ardent dog enthusiast will admit it's unlikely. Instead, Philip used his eyes and his body as communications.

Philip's behavior could be interpreted in three ways: one functional, one intentional, one conservative. The functional interpretation is this: the dog's gaze served as information for the person, whether the dog meant it to or not. The intentional: the dog did in fact mean it to: he looked because he knew the person was ignorant of the key's location. The conservative: the dog looked reflexively, since someone was recently over there where the key was.

The data do the interpreting. They show that the functional is definitely true: gaze did serve as information to the person nearby. But the intentional take is also true: the dog looked at the location of the key more often when the person in the room with them was ignorant where it was—as if meaning to inform the person with his gaze. That nixes the conservative interpretation. Philip seemed to be thinking about these crazy experimenters' minds.

This is but one dog—maybe a particularly astute one. Remember the begging experiment run with chimps and dogs? Unlike chimps, all the tested dogs immediately followed the knower's (non-blindfolded or bucketed person's) advice as to which box was baited with food. Hoorah for these dogs, who thus all found food inside. This looks good for the theory of dog mind: they acted as though thinking about the knowledge states of the strange people pointing in front of them. But after this seeming cognitive accomplishment, a strange thing happened. When run again and again on the same test, these dogs changed their strategies. They began to pick the guesser just about as often as the knower. Does this mean they were prescient and then grew dimwitted? Although dogs will do impressive convolutions for food, this doesn't make sense as an explanation. Perhaps it indicates that the first round was a fluke.

The best interpretation is that the dogs' performance on the task makes a methodological point. There may be other cues the dogs are using to make their decisions that are, to them, just as strong as the presence or absence of the guesser is to us. Consider, for instance, that all humans are on the whole highly knowledgeable about the sources of food, from a dog's point of view. We are regularly around food, we smell like food, we open and close a cold box filled with food all day long, and sometimes we even have food dribbling out of our pockets. This is such a well-learned feature of us that it might be hard to overturn on the basis of a few trials one afternoon. This hypothesis is borne out by the fact that the dogs did use the *people* to make their decision: they never chose a third box, unselected by either the guesser or the knower.

However we interpret the results, though, the dogs are not going out of their way to prove to us that they have a theory of mind. Of course, one of the difficulties of designing experiments for any animal is that, as the procedure grows more complicated in order to test for a very specific skill, it risks becoming an exceedingly strange scenario for the animal. One might suggest that massive confusion on the part of the subjects is not unreasonable. They are often thrust into situations that are bizarre: that are, in fact, intentionally unlike anything they've seen before. People appear with buckets over their heads; trials go on endlessly; it is in every way not normal. Dogs nonetheless sometimes manage to perform well at the tasks in front of them.

Still, their natural behavior—in a natural setting—is a better indication. What do dogs do without the peculiarities of baited and locked boxes and uncooperative humans to puzzle over? Their most representative behavior will appear in dealing naturally with other dogs or with humans. If it is socially helpful for a dog to consider what other dogs are thinking, the ability to do so may have evolved—and may still be visible in social interactions. This is why I spent a year watching dogs play: playing in living rooms and veterinary offices, down hallways and pathways, on beaches and in parks.

PLAYING INTO MIND

Pump appears in the corners of all the videos: in one, she hops nimbly to avoid collision with a dog approaching too fast— then pursues him as he rushes out of the frame. In another, she lies prone with another dog, feigning bites with open mouths. In a third she tries and fails to join two dogs in play; as they run off she is left wagging alone in the camera's eye.

I should correct myself: I was *lucky enough* to spend a year watching dogs play. What is called, appropriately, "rough-and-tumble" play between two competent, athletic dogs is a gymnastic marvel to witness. The playing dogs seem to give a perfunctory greeting to each other before they suddenly mutually attack, teeth bared; tumbling together in precarious free fall; jumping on and over each other; bodies bent and tangled. When they stop, suddenly, at a noise nearby, they may be the pictures of quiet. It takes only a look or a paw raised in the air to engage in their shared havoc again.

Play might seem just like *that thing dogs do,* but it has a very particular scientific definition. Animal play, science intones, is a voluntary activity incorporating exaggerated, repeated behaviors, extended or truncated in duration, varied in fortitude, and atypically combined; and using action patterns that have identifiable, more functional, roles in other contexts. We don't just define play this way to take the pleasure out of it: we define it to reliably recognize it. Play also has all the attributes of a good social interaction: coordination, turn taking, and, if necessary, self-handicapping—playing at the level of one's play partner. Each partner takes the abilities and behavior of the other into account.

The function of animal play is a bit of a puzzle. Most animal behaviors are described by how they function to improve the survival of the individual or species. The search for a function of play is paradoxical, as it looks like behavior which is clearly function*less:* at the end of play, no food has been gained, no territory secured, no mate wooed. Instead two dogs pantingly collapse on the ground and wag their tongues at each other. One might thus suggest that the function is *to have fun*—but this is frowned upon as a true function, because the risks are too great. Play takes a lot of energy, can cause injuries, and, in the wild, increases an animal's danger of predation. Play-fighting can escalate into true fights, causing not just injury but social upheaval. Its riskiness makes the case for a real, undiscovered function of play even more compelling: it must be terribly useful to play, if this behavior survived the evolutionary process. It might serve as practice: a context in which to hone physical and social skills. Strangely, though, studies have shown that play is not essential to adult proficiency at the skills practiced in play. Maybe play serves as training for unexpected events. It does seem that volatile and unpredictable play is deliberately sought. In humans, play is part of normal development—socially, physically, and cognitively. In dogs, it may be the result of having spare energy and time—and owners who live vicariously through their dogs' tumblings.

Play among dogs is particularly interesting because they play more than other canids, including wolves. And they play into adulthood, which is rare for most playing animals, including humans. Although we ritualize play into team sports and solo video game marathons, as sober adults we rarely spontaneously blindside and tackle our friends, tag them and run, or make faces at each other. The hobbling, slow-moving fifteen-year-old dog on the block looks warily at the enthusiasm of young puppies approaching him, but even he occasionally play-slaps and bites at a younger dog's legs in play.

In my study of dog play I shadowed dogs with a video camera rolling, and controlled my own delighted laughter at their fun long enough to record bouts of play, from a few seconds to many minutes long. After a few hours of this the fun stopped, the dogs would get packed into the backs of cars, and I would walk home, reflecting on the day. I'd sit down in front of my computer and play back the videos, at an extremely slow rate: slowed enough to see each frame—thirty of which fill a second—individually. Only at this speed could I really see what had happened in front of me. What I saw was not a repeat of the scene I'd witnessed at the park. At this speed I could see the mutual nods that preceded a chase. I saw the head-jockeying, open-mouth volleys that blurred into unrecognizability in real time. I could count how many bites it takes, over the course of two seconds, before a bitten dog responds; I could count how many seconds it takes for a paused bout to resume.

And, most important, I could look to see what behaviors dogs do, and when. Watching the play deconstructed into these subsecond moments enabled me to record a long catalog of the behaviors of each dog: a transcript of the play. I also noted their postures, their proximity to one another, and which way they were looking at every moment. Then, so deconstructed, the play could be reconstructed to see what behaviors match what postures.

In particular, I was interested in two kinds of behaviors: play signals and attention-getters. Attention-getters, as we've seen, are obvious things: they serve to get attention. Specifically, they are acts that alter the sensory experience of someone else—someone whose attention you're keen to have on you. They can be an interruption of the visual field, as when Pump suddenly puts her head between me and the book I'm holding. They can interrupt the auditory environment: a car's honk is so intended, and dogs' barks are so as well. If these methods fail, attention can be gotten by interacting physically: a hand on the shoulder; a paw on the lap; or, between dogs, a bump with the hip or a light bite on the rump. Clearly, many things we do are in some way attention-getting, but not every behavior is equally good at the task.

Calling your name out may be a way to catch your attention—but not if we are in Yankee Stadium in the bottom of the ninth. Then a more extreme method (and possibly an organist) would be necessary. Similarly, dogs' attention can be more or less easy to get. Between dogs, what I called an *in-your-face*—presenting oneself in front of, and very close to the face of, another dog—is effective at getting attention—but not if the dog is engaged in rollicking play with someone else. Then more forceful means are needed—thus explaining those dogs who circle a playing pair for minutes barking barking barking. (Better, perhaps, to interject some nice rump bites in there with the barking, if you are truly eager to break up the game.)

Play signals, the other behaviors, are requests for play or announcements of interest in playing: they could be translated as saying something like *Let's play* or *I want to play* or even *Ready? because I'm about to play with you.* What the specific words are is not as important as their functional effect: play signals are reliably used to begin and to continue play with others. They are a social requirement, not just a social nicety. Dogs typically play together rambunctiously and at a breakneck pace. Since they are doing all manner of actions that could easily be misinterpreted—biting each other on the face, mounting from behind or fore, tackling the legs out from under another dog—the playfulness of their actions has to be manifest.* If you fail to signal before biting, jumping on, hip-slamming, and standing over your playmate, you are not in fact playing; you are assaulting him. A

---

*Given the importance of regular visual assurances that the game is still a game, it is perhaps not surprising that successful three-way rough-and-tumble play is much rarer than play between two dogs. As with conversation, something is missed—a play signal here, an attention-getter there—when everyone is speaking at once. Typically, only dogs familiar with each other pull off threesomes.

bout wherein only one participant thinks it's play is no longer playful. All dog owners who walk their dogs among others know what then happens: a play bout becomes an attack. Without the play signal, a bite is a bite, worthy of rancor or retribution. With it, a bite is just part of the game.

Nearly every play bout begins with one of these signals. The quintessential signal is the *play bow,* in which the dog's body genuflects in front of a desired play partner. A dog bent on his forelegs, mouth open and relaxed, with his rump in the air and tail high and wagging is pulling out all the stops to induce someone to play. Even tailless, you can mimic this pose yourself; expect a response in kind, a friendly nip, or at least a second look. Two dogs who are regular playmates may use a bow shorthand: familiarity allows abbreviations in formality, just as between human acquaintances. Just as *How do you do?* became *Howdy,* the play bow can be shortened into the aforementioned *play slap,* the front legs clapping the ground at the beginning of the bow; the *open mouth display,* the mouth opened but without the teeth bared; or the *head bow,* a bobbing of the head with opened mouth. Even panting in quick bursts can be a signal to play.

It is how dogs might use these play-signal and attention-getting behaviors together that could reveal or refute that dogs have a theory of mind. In just the way the false belief task shows that some children are thinking about what other people know, and some are not, one's use of attention in communicating is meaningful. The key question I asked of my data of playing dogs was this: Did they communicate, using play signals, intentionally—with attention to the attention of their audience? And did they use attention-getters when they didn't have their play partner's attention? Just how were those bumps, barks, and bows of play used?

It's hard to give a good account of what's happened in a bout of play you have just watched. Sure, I could create a very simplistic story line between two dog protagonists—*Bailey and Darcy ran around together . . . Darcy chased Bailey and barked . . . they both bit at each other's faces . . . then they split*—but it glosses over the details, such as how often Darcy and Bailey self-hand-icapped, intentionally throwing themselves on the ground on their backs to be bitten, or using less force in a bite than they could. Whether they took turns in biting and being-bitten; chasing and being-chased. And, most critically, whether they signaled to each other when the signal could be seen and responded to—with play or by hightailing it out of there. For this, you need to look at the moments between the seconds.

What I found there was remarkable. These dogs play-signaled only at very particular times. They signaled reliably at the beginning of play—and always to a dog who was looking at them. Attention might be lost a dozen times in a typical play session. One dog gets distracted by a ripe smell underfoot; a third dog approaches the playing pair; an owner wanders away. What you might notice is simply a pause followed by a resumption of play. In fact, in these cases, a quick series of steps needs to be followed. For the play not to be permanently severed, the interested dog must regain his partner's attention and then ask him to play again. The dogs I observed also play-signaled when the play had paused and they wanted to resume the game—again, almost exclusively to dogs able to see the signal. In other words, they communicated intentionally, to an audience able to see them.

Even better, in many cases the record of where the dogs were looking revealed that a dog who had paused play was distracted—looking elsewhere, playing with someone else. One option for his erstwhile partner would be to play-bow madly, hoping to lure someone over to play. But more mindful would be just what they did: used an attention-getter before doing a bow. Importantly, they used attention-getters that *matched* the level of inattention of their playmates, showing they understood something about "attention." Even in the middle of play, they used mild attention-getters—such as an *in-your-face* or an *exaggerated retreat,* leaping backward while looking at the other dog—when their partner's attention was only mildly diverted. If a dog's desired playmate was just standing there staring at him, these attention-getters might indeed be enough to rouse him, as a wave *hello?* in front of a daydreaming friend. But when the other dog was very distracted, looking away or even playing with another dog, they used assertive attention-getters—bites, bumps, and barks. In these cases, that mild *hello?* would not do. Instead of using a brute-force method of trying to get attention by any means necessary, they chose types of attention-getters that were just sufficient, but not superfluous, to get the desired attention. This was truly sensitive behavior on the part of the players.

Only after these attention-getters were successful did the dogs signal their interest in playing. In other words, they were using an order of operations: get attention first, then send an invitation to rumble.

This is just what good theorists-of-mind do: think about their audience's state of attention and only talk to those who can hear and understand them. The dogs' behavior looks tantalizingly close to a display of theory of mind. But there's reason to believe that their ability is different than ours. For one thing, in both the experiments and my play study, not all dogs acted equally mindfully. Some dogs are oblivious in their attention-getting. They bark, get no response—and then bark and bark and bark and bark. Others use attention-getters when attention has already been gotten, or play signals when play has already been signaled. The statistics show that most dogs act mindfully, but there are plenty of exceptions. We can't tell yet whether they are just the underperformers or whether they indicate that the species has an incomplete understanding.

It may be a little of both. Rather than contemplating the mind behind the dog, most dogs are likely to simply interact. Their skill at using attention and play signals hints that they may have a *rudimentary* theory of mind: knowing that there is some mediating element between other dogs and their actions. A rudimentary theory of mind is like having passable social skills. It helps you play better with others to think about their perspective. And however simple this skill may be, it may be part of an inchoate system of fairness among dogs. Perspective-taking underlies our agreement to a code of conduct between humans that is jointly beneficial. Watching play, I noticed that dogs who violated the implicit rules for attention-getting and play-signaling—simply barging in on others' play without following the proper, mindful procedures, say—were shunned as playmates.*

Does this mean that your dog is aware of and interested in

---

*Another indication of the dog's perception of fairness comes from a new experiment demonstrating that dogs who see another dog getting a reward for doing an act—shaking a paw on command—but who do not themselves get rewarded for the same act eventually refuse to shake anymore. (No rewarded dog was moved by the clear injustice of the situation to share his earned bounty with his unlucky partner, though . . .)

what's on your mind right now? No. Does it mean that he might realize that your behavior reflects what's on your mind? Yes. Used to communicate with us, this is a large part of dogs' seeming humanity. Sometimes it is even used in nefarious ways only too human.

WHAT HAPPENED TO THE CHIHUAHUA

We can now revisit the wolfhound and Chihuahua we met at the start of this book. Their hillside encounter is no less remarkable now, but it does perfectly encapsulate the flexibility and variety of behaviors of the species. The explanation for that play begins in the history of their social ancestors, the wolves; it is apparent in the hours of socializing between humans and dogs; in the years of domestication; in the dialogues of speech and behavior between us. It is explicable in the sensorium of the dog: the information he gets from his nose, what his eyes take in. It is in the capacity of dogs to reflect on themselves; it is explained in their different, parallel universe.

And it is in the particular signals they use with each other. The wolfhound's high-rumped approach: the play bow, an invitation to a game—making perfectly clear his ardent intent to play with, not eat, the little dog. In return, the Chihuahua bowed: accepting the offer. In the language of dogs that is enough to see each other as equals in play. Their disparate sizes aren't irrelevant—and this explains the hound's drop to the ground: he handicaps himself. By putting himself at the little dog's height—taking the Chihuahua's point of view—and exposing himself to her attacks, he levels the playing field.

They endure jostling body on body. Bodies in full contact is a reasonable social distance for dogs. They bite with impunity:

every bite is matched, or explained with a play signal—and every bite is restrained. When the hound hits the little dog too hard, sending her scurrying backward, she could for a moment be seen as small, fleeing prey. But the difference between dogs and wolves is that dogs can put aside their predatory instincts. Instead the hound takes back that swipe with an apologetic play slap, a milder version of the bow. It works: she rushes right back into his face.

Finally, when the hound is pulled up and away by his owner, the Chihuahua tosses a bark to her departing playmate. Had we kept watching them, had he turned around, we might have seen her open her mouth or leap a tiny leap—calling out in the hopes of continuing the game with her giant friend.

NON-HUMAN

The study of dogs' cognitive abilities emerged from a context of comparative psychology, which by definition aims to compare animals' abilities with those of humans. The exercise often winds up splitting hairs: they communicate—but not with all the elements of human language; they learn, imitate, and deceive—but not *in the way* that we do. The more we learn of animals' abilities, the finer we have to split the hair to maintain a dividing line between humans and animals. Still, it is interesting to note that we seem to be the only species spending any time studying other species—or, at least, reading or writing books about them. It is not necessarily to the dogs' discredit that they do not.

What is revealing is how dogs perform on tasks that measure social abilities we thought only human beings had. The results, whether serving to show how alike or unalike dogs are to or from us, have relevance in our relationships with our dogs. When considering what we ask of them and what we should expect from them, understanding their differences from us will serve us well. Science's effort to find distinctions illustrates more than anything else the one true distinction: our drive to affirm our superiority—to make comparisons and judge differences. Dogs, noble minds, do not do this. Thank goodness.

Inside of a Dog

Her personality is unmistakable and omnipresent: in her reluctance to climb the steep steps out of the park—but then forging ahead of me strongly and gamely; in her great spasms of running and scent rolling of younger days; in her delight at my return from a long trip—but not dwelling on it; in her checking back for me on our walks but also always keeping a few paces apart. For a dog who is in fact wholly dependent on me, she is incredibly independent: her personality is forged not just in interaction with me, but in the times wandering outside without me, in exploring her space alone. She has her own pace of life.

Despite the wealth of scientific information about the dog—about how they see, smell, hear, look, learn—there are places science doesn't travel. It perplexes me that some of the questions I have most often been asked about dogs, and that I have about my own dog, are not addressed by research. On matters of personality, personal experience, emotions, and simply *what they think about,* science is quiet. Still, the accumulation of data about dogs provides a good foothold from which to extrapolate and reach toward answers to those questions.

The questions are typically of two kinds: *What does the dog know?* and *What is it like to be a dog?* So first we will ask what dogs know about things of human concern. Then we can further imagine the experiences—the umwelten—of the creatures who have this knowledge.

I

WHAT A DOG KNOWS

Claims about what dogs know are made constantly. Oddly, they tend to cluster around the academic and the ridiculous. The former prompts researchers to ask if a dog knows how, for instance, to count sums. In one experiment the dogs looked longer— evincing surprise—when there were either more or fewer biscuits revealed behind a screen than they had been shown being hidden there one by one—indicating that they were keeping track of *number* and noticing when there was a discrepancy. Ta-da: counting dogs.

The other kind of claims are the far-flung: that dogs have ethics, rationality, a metaphysics. I admit to entertaining the notion more than once that my own dog seems to act ironically (whether or not she intends to).* One ancient philosopher maintained that dogs understand disjunctive syllogisms. As evidence,

---

*As when she spent a quarter hour digging a hole in which to drop a treasured rawhide chew, but in digging actually created more of a pile than a hole: the result being that the rawhide was actually not stashed secretly away, but proudly and conspicuously displayed (itself probably the result of an imperfectly-honed caching instinct). In like manner, one might wonder if she experiences it as ironic (or as magic) when I make a show of unfurling my fingers in front of her and the treat I had in my palm is missing.

he gave the observation that in tracking an animal to a branching path, dogs can deduce that if the animal is neither down the first nor second of three trails, they realize, even without scent, that it must be down the third.*

Starting with an interest in math or metaphysics and working downward does not get us very far in understanding dogs. But start with their snuffling approach of the world, their striking attention to humans, and knowledge of the various means by which dogs learn about the world—and we might be able to learn what they know. In particular, we might approach an answer to whether they experience life as we do: whether they think about the world as we do. We mind our own autobiographical journeys through life, managing daily affairs, plotting future revolutions, fearing death, and trying to do good. What do dogs know about time, about themselves, about right and wrong, about emergencies, emotions, and death? By defining and deconstructing these notions—making them scientifically examinable—we can begin to answer.

Dog days (About time)

Back home, Pump gives me a perfunctory greeting, executes an unlikely pirouette, and then races off. Over the course of the day she has located all the biscuits I left around the house for her, and has waited until now to consume them, gobbling from the one balanced on the chair's edge to the one on the doorknob to the tricky one on a towering pile of books, which she delicately plucks off and spirits away.

*This could be another way of accounting for Rico's ability to pick the toy with the unfamiliar name out of a pile of toys: he selected the toy that he did not recognize.

Animals exist in time, they use time; but do they experience time? Surely they do. At some level there is no difference between existing in time and experiencing time: time must be perceived to be used. What many people mean, I suspect, in asking whether animals experience time is, Do animals have the same feelings about time that we do? Can a dog sense the passage of a day? And, critically, are dogs bored all day, at home alone?

Dogs have plenty of experience of the Day, if no word *day* to call it. We are the first source of their knowledge of days: we organize the dog's day in parallel with ours, providing landmarks and surrounding them with ritual. For instance, we provide all sorts of cues about when the dog's mealtime is. We head for the kitchen or pantry. It may be our mealtime, too, so we begin to unload the refrigerator, wafting food smells about, and making a racket with pots and plates. If we glance at the dog and coo a little, any remaining ambiguity is erased. And dogs are naturally habitual, sensitive to activities that recur. They form preferences—places to eat, to sleep, to safely pee—and notice preferences of yours.

But in addition to all those visible and olfactory cues, does the dog naturally know that it is dinnertime? I know owners who insist they can set the clock by their dog. When he moves to the door, it's precisely the time to go out; when he moves to the kitchen, sure enough, it's time to be fed. Imagine removing all the cues the dog has about the time of day: all of your movements, any environmental sounds, even light and dark. The dog still knows when it's time to eat.

The first explanation is that dogs wear an actual clock—though internally. It is in the so-called *pacemaker* of their brain, which regulates the activities of other cells of the body through the day. For a few decades neuroscientists have known that circadian rhythms, the sleep and alertness cycles that we experience every day, are controlled by a part of the brain in the hypothalamus called the SCN (suprachiasmatic nucleus). Not only humans have an SCN: so do rats, pigeons, dogs—every animal, including insects, with a complex nervous system. These neurons and others in the hypothalamus work together to coordinate daily wakefulness, hunger, and sleep.* Deprived entirely of cycles of light and dark, we would all still go through circadian cycles; without the sun it takes just over twenty-four hours to complete a biological day.

> This morning I heard her barking in her sleep—the muffled, jowl-puffing bark of dreaming. Oh, does she dream. I love her dream-barks, falsely severe, often accompanied by twitching feet or lips curled into a teeth-baring growl. Watch long enough and I'll see her eyes dancing, the periodic clenches of her jaw, hear her tiny whimpers. The best dreams inspire tail-wags— huge thumps of delight that wake herself and me.

We humans experience the day according to our ideas about what typically or ideally will happen throughout it—what meals, work, play, conversation, sex, commuting, naps—and also according to the cycle of our circadian rhythms. Given our attention to the former, though, we sometimes hardly notice that our bodies are charting a regular course through the day. That midafternoon sleepiness, the difficulty in rising at five in the

---

*With age, dogs sleep more but enter paradoxical—REM—sleep less than in youth. Scientists have theories but no final explanation for why dogs dream—and they dream vividly, if their eye fluttering, claw curling, tail twitching, and yelping in sleep is any indication. As in humans, one theory names dreams the accidental result of paradoxical sleep, which itself is a time of bodily restoration; alternatively, dreams might function as a time to practice, in the safety of one's imagination, future social interactions and physical feats or to review interactions and feats past.

morning—both are due to our activities clashing with our circadian rhythms. Take away some of those human expectations and you've got the dog's experience: the bodily feelings of the passage of the day. In fact, without the societal expectations to distract them, they may be more attuned to the rhythms of their body telling them when to rise and when to eat. As per their pacemaker, they are most active as dark gives way to dawn, and markedly reduce their activity in the afternoon, with a burst of energy in the evening. With nothing else to do—no papers to shuffle, no meetings to attend—dogs nap straight through that afternoon slowdown.

Even without regular mealtimes the body goes through feeding-related cycles. Right before it is time to eat, animals tend to be more active—running about, licking, salivating—in anticipation of food. We see this food-sense when a dog pursues us relentlessly with panting mouth and appealing eyes. Eventually we figure out it is time to feed the dog.

So in fact one can set the clock by the dog's belly. And, even more impressive, dogs maintain a clock operated by other mechanisms not yet fully understood, which seem to read the day's air. Our local environment—the air in the room we are in—indicates (if we have the right indicator) where we are in the day. Although we do not typically sense it, it is just the sort of thing a dog might notice. If we attend carefully, we might notice the gross changes of the day: the cool at the moment the sun sets, or the time of day registered in the amount of light streaming in the window—but the day's changes are infinitely more subtle than this. With sensitive machinery, researchers can detect the gentle air currents that form as a summer's day ends: warmed air pulled up along the inner walls creeps across the ceiling, spilling into the center of the room and falling along the outer walls. This is no breeze, nor even a noticeable puff or waft. Yet the sensitive machinery that is the dog evidently detects this slow, inevitable flow of air, perhaps with the help of their whiskers, well positioned to register the direction of any scent on the air. We know they can detect it because they can also be fooled: brought into a room that was warmed, a dog trained to follow a scent trail may search first by the windows when the track is really closer to the room's interior.

She is patient. How she waits for me. She waits as I duck into the local grocery store: looking plaintively, then settling down. She waits at home, warming the bed, the chair, the spot by the door, for me to return. She waits for me to finish up what I'm doing before we go outside; for me to finish talking with someone during our walk; for me to figure out when she is hungry. She waited for me to finally realize where she liked to be rubbed. And for me to finally begin to figure her out. Thanks for waiting, kiddo.

Dogs have not been tested on their ability to detect a specific length of time; but bumblebees have. In one study, bees were trained to wait for a fixed time interval before sticking a proboscis through a tiny hole for a bit of sugar. Whatever the interval, they learned to restrain themselves for just that long . . . and then no longer. When you're a bee waiting for sugar water, a half minute is a long time to wait. But they patiently tapped their many feet and did so. Other well-experimented-on animals— rats and pigeons—do the same: measuring time.

It is probable that your dog knows just how long a day is. But if so, a horrible thought occurs: Mustn't dogs be terribly bored enduring that day all alone at home? How can we tell if a dog is bored? Like other concepts whose applicability to dogs we are curious about, we first need to get a handle on what boredom looks like. Any child will tell you when he is bored, but dogs don't—at least, not verbally.

Boredom is rarely discussed in the non-human scientific literature, because it is one of the classes of words whose application to animals is thought suspect. "Man is the only animal who can be bored," the social psychologist Erich Fromm declared; dogs should be so lucky. Human boredom is rarely the subject of scientific scrutiny, either, perhaps because it is seen simply as a part of the experience of life, not as a pathology to scrutinize. Its very familiarity gives us a way to define it: we experience it as a profound ennui, as an utter lack of interest. And we can recognize it in others: in their flagging energy, in an uptick in repetitive movements and a decline in all other activities, and in rapidly waning attention.

With this definition, the subjective becomes objectively identifiable, in dogs as well as humans. Flagging energy and reduced activity are simple to recognize: less moving and more lying and sitting. Attention may wane straight into protracted bouts of sleep. Repetitive movements include stereotyped (aimlessly and endlessly repeated) or self-directed behaviors. We twiddle our thumbs when bored; we pace. Animals kept in barren zoo enclosures often pace madly—and, thumbless, have twiddle-equivalents: licking or chewing skin or fur obsessively and constantly, pulling out their own feathers, rubbing their ears or face, rocking back and forth.

So is your dog bored? If you return home to find apparently restless socks, shoes, or underwear that have magically migrated some small distance from where you left them, or straggled bite-sized reminders of what you threw in the garbage yesterday—the answer is both *Yes,* your dog was bored, and *No,* at least not during one manic hour of chewing. Imagine a child complaining, *There's nothing to do:* that is just the case for most dogs left alone. Left without anything to do, they will find something. Your solution, for the sake of your dog's mental health, and for the sake of your socks, is as simple as leaving something for them to do.

Even if you return to find the house a bit unkempt, a warm depression on the forbidden couch cushion, what is also reliable is that the dog is still alive and usually looks well. We get away with leaving them, with boring them, because they generally adapt to their situations without much complaint. In fact, dogs take comfort in habit, in reliable occurrences, just as we might. If so, then their boredom may be tempered by resignation to the familiar. And they may even know how long they typically need to stay in the suspended animation of waiting at home for you. It is one reason why your dog may be waggily waiting at the door even when you try to quietly sneak in at the workday's end. And it is why I leave more treats hidden around the apartment the longer I will be gone. I'm telling Pump I'll be away—and leaving something to mind the time.

The inner dog (About themselves)

The best scientific tool proposed to determine if dogs think about themselves—if they have a sense of self—is a simple one: the mirror. One day the primatologist Gordon Gallup pondered his reflection while shaving and wondered if the chimpanzees he studied would ponder their reflections in mirrors, too. Certainly using a mirror for self-examination—smoothing a shirt over belly, patting down a wayward hair, testing a coy smile—is a display of our own self-awareness. And before we are self-aware, as young children, we do not use mirrors as adults do. A short time before children pass theory-of-mind tests, they begin to consider their mirror images.

Gallup promptly placed a full-length mirror outside his chimpanzees' cages and watched what they did. They all did the same thing first: they threatened and tried to attack the mirror. Suddenly, it seemed, there was another chimp right outside their cage; this must be addressed at once. Despite the no doubt confusing result—the mirror image seemed to attack back, only for the affair to resolve without ado—their first days with the mirrors were full with social displays toward this new, glaring chimpanzee. After a few days, though, the chimpanzees seemed to come to a realization. Gallup watched as his chimps approached the mirrors and began to use them to examine their own visages and bodies: picking at their teeth, blowing bubbles, making faces toward their mirror image. They were especially interested in parts of their bodies that are ordinarily visually inaccessible: the mouth, the rump, up the nostrils. To be sure that they were thinking about the mirror images as *themselves,* Gallup devised a "mark" test: he inconspicuously applied a prominent dab of red ink to the head of the chimps. These first subjects in this test needed to be anesthetized to apply the mark; later researchers would affix the mark while doing ordinary grooming or medical care of their animals. When the marked chimps again stood in front of the mirrors, they saw a red-tagged chimp—and they touched the spot on their own heads, bringing their hands down to examine the ink with their mouths. They passed the test.

There is considerable debate about whether this indicates that chimpanzees are thinking about themselves, have a concept of self, recognize themselves, are self-aware, or none of the above*—especially since it would be disruptive of our ideas about animals to suddenly grant them self-awareness. But the mirror tests have continued alongside the debate, and to this date dolphins (by moving their bodies to explore the mark) and at least one elephant (using her trunk) have passed the test; monkeys have not. And dogs? Dogs have not been shown to pass the test. They never examine themselves in the mirror. Instead they behave more like monkeys do: they sometimes look at the mirrors as though it were another animal, and sometimes look at it idly. In some cases, dogs will use mirrors to get information about the world: to see you tiptoeing up behind them, for instance. But they don't seem to see the mirror as an image of themselves.

There are a few explanations why dogs might behave this way. The dogs may indeed not have any sense of self—thus no sense of who that handsome dog in the mirror might be. But as the debate over this test indicates, it is not universally accepted as a conclusive test of self-awareness; thus neither can it be a conclusive determination of lack of self-awareness. Another possible explanation for the dogs' behavior is that the lack of other cues—specifically olfactory cues—coming from the mirror image leads dogs to lose interest in investigating it. Some fantastical odor-mirror that wafts the dog's own scent while reflecting the dog's own image would be a better medium for this test. Another issue is that the test is predicated on a specific kind of curiosity about

---

*When animals pass the test, skeptics highlight the logical fallacy of the conclusion: that self-aware humans use a mirror to examine ourselves does not imply that using a mirror requires self-awareness. When animals fail the test, the debate goes the other way: there is no good evolutionary reason why animals *should* examine something non-irritating on their heads, even if they recognized themselves. In either event, the mirror test continues to be the best test thus far developed for self-awareness, and one that uses simple equipment to boot.

oneself: one that leads humans to examine what is new on our own bodies. Dogs may be less interested in what is visually new than what is tactually new: they notice strange sensations and pursue them with nibbling mouth or scratching paw. A dog is not curious why the tip of his black tail is white, or what the color of his new leash is. The mark needs to be noticeable, and also worth noting.

Even so, there are other dog behaviors suggestive of their self-knowledge. In most actions, dogs do not grossly misestimate their abilities. They surprise themselves by jumping into water after ducks—only to find that they are natural swimmers. They surprise us by leaping to scale a fence—which they may in fact be able to clear. On the other hand, one regularly hears that dogs *don't* know a very basic fact about themselves: how big they are. Small dogs strut up to enormous dogs: their owners proclaim that their dogs "think they're big." Some big-dog owners who endure lap sitting likewise assert that their dogs "think they're small." In both cases, the dogs' accompanying behaviors lend more credibility to the notion that they *do* know their sizes: the small dog is compensating for his small size by trumpeting his other qualities extra loudly; the large dog raised with a lap to sit on continues with this close contact just as long as he is tolerated, and then finds a large-dog-sized pillow to sit on elsewhere.

Both small and large dog are tacitly acknowledging an understanding of their own size. It might seem unlikely that this means they are thinking about the categories *big* or *small*. But look at how they act on objects in the world. Some dogs will attempt to pick up a felled tree, but most dogs with stick-carrying habits will choose similarly sized sticks at every opportunity, as though they have gauged what can be picked up and held in their mouths. From then on, all sticks in the path of a searching dog are quickly assessed: too big? too thick? not thick enough?

Further suggestive evidence that dogs know their size comes from their rough-and-tumble play. One of the most characteristic features of dog play is that socialized dogs can, by and large, play with almost any other socialized dog. This includes the pug who leaps onto the hocks of the mastiff, reaching his knee. As we've seen, big dogs know how to, and often do, moderate the force of their play to smaller playmates. They can withhold their fiercest bites, jump halfheartedly, bump into their more fragile playmates more gently. They might willingly expose themselves to attack. Some of the largest dogs regularly flop themselves on the ground, revealing their bellies for their smaller playmates to maul for a while—what I called a *self-takedown*. Older, learned dogs adjust their play styles to puppies, who don't yet know the rules of play.

Play between dogs of mismatched statures often does not last long, but it is usually an owner, not a dog, who moves to stop it. Most socialized dogs are considerably better at reading each other's intent and abilities than we are. They settle most misunderstandings before owners even see them. It's not the size or the breed that matters; it's the way they talk to each other.

Working dogs provide another glimpse into what dogs know about themselves. Sheepdogs, raised from their first weeks of life with sheep, do not grow up to act like sheep. They do not bleat or scream, chew their cud, aggressively head-butt, nor suckle from the ewe, as sheep do. Their cohabitation leads dogs to interact socially with sheep—using social behaviors characteristic of dogs. Those who study sheepdogs observe, for instance, that dogs will growl at sheep. Growling is a dog communication: the dog is treating the sheep more like a dog than like a possible meal. These dogs' only fault is to overgeneralize: not only are they clear on their own identity, in some sense—they also think that everyone else is a dog, too. One could call this foible very human: they talk to sheep as though they were dogs, just as we talk to dogs as though they were humans.

Between play bouts, stick-retrieving, and sheepherding, do dogs sit around thinking, *My, but I'm a fine medium-sized dog, aren't I?* Certainly not: such continued reflection on size or status or appearance is peculiarly human beings' lot. But dogs do act with knowledge of themselves, in contexts where such knowledge is useful. They respect (for the most part) the limits of their physical abilities, and will look pleadingly at you when you ask them to leap a too-high fence. A dog will hop discreetly around a pile of his own defecation encountered on the ground: he recognizes the smell as *his*. If the dog is reflecting on himself, one might wonder if he thinks about himself in the past—or in the future: if he is quietly writing his autobiography in his head.

Dog years* (About their past and future)

As we round the corner Pump stops in her tracks. She moves as if to sniff something a half-step back; I slow to indulge her; and

---

*I do not know if the origin of the myth of *dog years*—that dogs live the equivalent of seven years for every one of our years—has ever been cracked. I'd guess that it is a backward extrapolation from the length of the expected lifetime of humans (seventy+ years) to the expected lifetime of dogs (ten to fifteen). The analogy is more convenient than it is true. There is no real life-length equivalence except that we both are born and die. Dogs develop at lightning speed, walking and eating on their own in their first two months; human infants take over a year. By a year, most dogs are accomplished social actors, able to navigate dog and human worlds easily. The average child might be there by four or five. Then dog development slows, while human development skyrockets. If committed to the comparison, one could make a case for a sliding scale ratio: around 10 to 1 in their first two years, then diminishing to more like 2 to 1 in their last years. But the truly committed should consider the critical-period windows, the performance on cognitive tests, the diminishment of sensory capacities with age, and the lifespans of different breeds in their calculus.

she darts back around the corner. There are still twelve blocks, a brief park, a water fountain, and a right turn until we get there, but she knows this walk. She'd been glancing up at me for blocks, and with that final turn, it's confirmed. We're going to the vet.

Psychologists report that those people with the most prodigious memories—able to flawlessly recite a string of hundreds of random numbers read to them once, as well as identifying every moment the reader blinked, swallowed, or scratched his head—are sometimes the most tortured by what they recall. The complement of remembering so thoroughly can be the strange inability to forget anything at all. Every event, every detail, piles on the garbage heaps that are their memories.

The overflowing garbage, collector of the day's past, is more than a little evocative when considering the memory of a dog. For if anything is on the dog's mind, it is that wonderful, odoriferous pile that we teasingly preserve in our kitchens, off-limits to the dog as a special form of torture. In that pile go the leavings of so many dinners, the extra-rank cheese that was discovered in the back of the fridge, clothes that have smelled too much for too long to be worn. Everything goes there but nothing is organized.

Is the dog's memory like this? At some level, it just might be. There is clear evidence that dogs remember. Your dog plainly recognizes you on your return home. Every owner knows that their dog won't forget where that favored toy was left, or what time dinner is supposed to be delivered. He can forge a shortcut en route to the park; remember the good peeing posts and quiet squatting sites; identify dog friends and foes at a glance and a sniff.

However, the reason we even pose the question "Do dogs remember?" is that there is more to our memory than keeping track of valued items, familiar faces, and places we've been. There is a personal thread running through our memories: the felt experience of one's own past, tinged with the anticipation of one's own future. So the question becomes whether the dog has a subjective experience of his own memories in the way that we do—whether he thinks about the events of his life reflexively, as *his* events in *his* life.

Though usually skeptical and reserved in their pronounce-ments, scientists often implicitly act as though dogs have memo-ries just like ours. Dogs have long been used as models for the study of the human brain. Some of what we know about the diminishing of memory with age comes from tests on the dimin-ishment of the beagle's memory with age. Dogs have a short-term, "working" memory that is assumed to function just as the psychology primers teach that human memory works. Which is to say: At any moment, we are more likely to remember just those things that we bring a "spotlight" of attention to. Not everything that is happening will be remembered. Only those things that we repeat and rehearse for later recollection will get stored as longer-term memories. And if a lot is happening at once, we're bound to remember only some of it—the first and last things sticking best. The dog's memory works the same way.

There is a limitation to the sameness. Language marks the difference. One reason why as adults we don't have many— arguably any—true memories of life before our third birthdays is that we were not skilled language users at the time, able to frame, ponder, and store away our experiences. It might be the case that while we can have physical, bodily memories of events, people, even thoughts and moods, what we mean by "memories" is something facilitated only by the advent of linguistic compe-tence. If that's the case, then dogs, like infants, don't have that kind of memory.

But dogs certainly remember a large amount: they remember their owners, their homes, the place they walk. They remember innumerable other dogs, they know about rain and snow after experiencing them once; they remember where to find a good smell and where to find a good stick. They know when we can't see what they are doing; they remember what made us mad last time they chewed it up; they know when they are allowed on the bed and when they are forbidden from it. They only know these things because they have learned them—and learning is just memory of associations or events over time.

Back, then, to the matter of the autobiographical memory. In many ways, dogs act as if they think about their memories as the personal story of their life. They sometimes act as though they are thinking about their own future. Unless sick or asleep, there was usually nothing that could stop Pump from eating dog biscuits—and yet she often refrained when home alone, opting to wait for my return. Even when accompanied, dogs regularly hide bones and squirrel away other favored treats; a toy may be abandoned outside with seeming insouciance only to be beelined-for the next week. Their actions can often be traced to events of their own past. They remember and avoid ground that was rough underfoot, dogs who turned suddenly gruff, people who acted erratically or cruelly. And they evince familiarity with creatures and objects they encounter repeatedly. Besides their quick recognition of their new owners, young dogs come to know their owners' visitors over time. They play best, and with the least ceremony, with those dogs they have known the longest—as though they are stamped together. These longtime playmates need not use elaborate play signals with each other: they use their own shorthand, signals abbreviated into mere flashes, before fully engaging.*

It is somewhat dispiriting to find that our knowledge about a dog's autobiographical sense has not advanced beyond Snoopy's affirmation half a century ago, "Yesterday I was a dog. Today I'm a dog. Tomorrow I'll probably still be a dog." No experimental study has specifically tested the dog's considerations of his own past or future. But a few studies with other animals examine part of what might be considered their autobiographical consciousness. For instance, a test run on the Western scrub-jay, a bird that naturally caches food for later consumption, has shown what in humans would be called willpower. If I'm hankering for chocolate-chip cookies, and someone gives me a bag of chocolate-chip cookies, it is extremely unlikely that I would put them away until the next day. The jays were taught that when given a preferred food—their chocolate-chip cookie equivalent—they would not be given food on the subsequent morning. Despite what we can presume is a strong interest in eating the food straightaway, they saved some and consumed it the next day. And me, without my cookies.

We might ask whether dogs act similarly. If prevented from

---

*This is similar to what has been called *ontogenetic ritualization:* the co-shaping by individuals of a behavior over time, until even the very initial part of the behavior carries meaning for them. In humans, an eyebrow-raise from one friend to another can take the place of a spoken commentary; as we've seen, among dogs a quick head-raise might replace an entire play bow.

eating in the mornings, does your dog begin to stash food the night before? If so, that would be suggestive evidence that they can plan for the future. As we know from finding uneaten unidentifiables in refrigerated takeout containers, not all saved food is equally good over time. If your dog buries a bone in the dirt or in the corner of the couch each month for three months, does he remember which is the oldest, the foulest, and which is the freshest? Putting aside any overpowering odors emanating from your couch, it is not likely. If we consider the dog's environment, it is apparent that they simply do not need to use time in this way, as they, unlike scrub-jays, are provided with a regular supply of food. In addition, discriminating food by its expiration date, or saving food for later when you're hungry now, may be a difficult task for an animal descended from opportunistic feeders, who eat as much as they can when food is available, then endure long stretches of fasting when food is not. Some suggest, reasonably, that dogs' bone-burying behavior is tied to an ancestral urge to stash some food aside for the lean times.* Evidence that a dog can distinguish the freshest bone from the one that has rotted—or leaves some aside just to enjoy it later—would bear this out. It is more likely that most of the time dogs are not thinking about time when they are thinking about food. A bone is a bone is a bone, buried or in the mouth.

On the other hand, a dearth of evidence verifying dogs' time-telling with bones does not mean that dogs do not distinguish past from present from future. When encountering a dog who had once—but only once—been aggressive, a dog will first

---

*Which some wolves instinctively do: even as young cubs they burrow their noses into a patch of land, drop a bone, nose-burrow some more, then proudly leave their poor excuse of a hole with a bone obviously visible. As adults they refine the behavior and do retrieve cached food—although there is no data about whether the retrieval is time-sensitive.

be wary and gradually, with time, grow more emboldened. And dogs certainly anticipate what is in their near future: with growing excitement on beginning the walk that leads to the dog food store; or anxiety at the car ride that suggests a visit to the veterinarian.

Some thinkers treat the dog as having no past: as enviably ahistorical, happy because they cannot remember. But it is clear that they are happy even despite remembering. We don't yet know if there is an "I" there behind the dog's eyes—a sense of self, of being a dog. Perhaps there need only be a continuous teller for the autobiography to be written. In that case, they are writing it right now in front of you.

Good dog (About right and wrong)

When Pump was a young dog, a common scene in our household went like this: I turn my back or go into another room. Milliseconds later, Pumpernickel has her nose at the kitchen trash can, peering in for good bits. If I return and catch her in this vulnerable spot, she immediately pulls her nose out of the can, her ears and tail drop, and she wags excitedly, slinking away. Caught.

When researchers asked a sample of dog owners what kinds of things dogs know or understand about our world, the owners most frequently claimed that dogs know when they have done something wrong: that dogs have knowledge of a kind of category of *things one must never, ever do.* These days that category includes things like tearing into the garbage, devouring footwear, and snatching just-cooked food off the kitchen counter. The punishment in our enlightened age is, one hopes, not terribly severe:

a stern word; a frown and a stamped foot. It was not always so: in the Middle Ages and earlier, dogs and other animals were brutally punished for misdeeds, from the "progressive mutilation" of the ears, feet, and on to the tail of a dog in correspondence with the number of people he had bitten, to the capital punishment, after legal trial and conviction, of a dog for homicide;* to earlier, in Rome, the ritual crucifixion of a dog on every anniversary of the evening the Gauls attacked the capital and a dog failed to warn of their approach.

The guilty look of a dog responsible for lesser trespasses is well-known to anyone who has caught a dog in Pump's pose, with her snout deeply plunged in the trash can, or discovered with bits of stuffing in his mouth and surrounded by tufts of what had until recently been the innards of the couch. Ears pulled back and pressed down against the head, tail wagging in quick time and tucked between the legs, and trying to sneak out of the room, the dog gives every appearance of realizing that he's been caught red-pawed.

The empirical question this raises is not whether this guilty look reliably occurs in such settings: it does. Instead, the question is what it is, exactly, about those settings that prompts the look. It may in fact be guilt—or it may be something else: the excitement of sniffing the trash, a reaction to being discovered, or anticipation of the unhappy, loud noises her owner tends to make when encountering trash out of its can.

Can dogs know right from wrong? Do they know that *this particular action* is clearly, maddeningly, wrong? A few years

*The medieval policy seems ridiculous to presume that dogs merit lawful consideration. It may seem equally ridiculous that our modern policy presumes that dogs do not: we still kill dogs who mortally wound a human—but now we call the dogs "dangerous" and do not bother to put them on trial (though their owners might be tried).

back, a Doberman employed to guard an expensive teddy bear collection (including Elvis Presley's favorite bear) was discovered in the morning with the devastation of hundreds of maimed, mauled, and beheaded teddies around him. His look, captured in news photos, was not of a dog who thought he had done wrong.

It would seem to defy reason if the mechanism behind the guilty or the defiant look were the same as ours. After all, right and wrong are concepts that we humans have by virtue of being raised in a culture that has defined such things. Excepting young children and psychotics, every person winds up knowing right from wrong. We grow up in a world of oughts and oughtn'ts, learning some rules for conduct explicitly and others by a kind of observational osmosis.

But consider how we know that other people know right from wrong when they cannot tell us so. A two-year-old sidles up to a table, gropes toward an expensive vase, and knocks it over, shattering it. Does the child know that it is wrong to break things that belong to other people? This might be an occasion on which, given the probable explosive reaction from any adults in the vicinity, she begins to learn. But at age two, she does not yet understand the concepts: she did not maliciously destroy the vase. Instead, she is an ordinary two-year-old who is clumsily trying to master moving her own body. We get an indication of her intent by watching what she did before and after the vase fell. Did she head directly for the vase and act to push it over? Or was she reaching for the vase and was uncoordinated in doing so? After it fell, did she evince surprise? Or did she look, well, satisfied?

Essentially the same method can be applied to dogs by allowing them to break expensive vases and watching how they react. I designed an experiment to determine if those guilty looks come from being guilty or from one of the something-elses.

Though my method is experimental, the setting is ordinary, so as to best capture the animals' natural behavior: in the "wild" of their own homes. To qualify for subjecthood, dogs had to have been exposed to an owner's disallowance—for instance, by the owner pointing at an object to be left alone and loudly stating *No!**—and must know to therefore leave it be.

In the place of expensive vases, I use highly desirable treats—a bit of a biscuit, a cube of cheese—that will not be shattered, but will be expressly forbidden. Given that the claim being tested is that a dog knows that engaging in a behavior that has been disallowed by the owner is wrong, I designed this experiment to provide an opportunity to do that very behavior. In this case, the owner is asked to bring the dog's attention to the treat and then clearly tell the dog not to eat it. The treat is placed in an enticingly available spot. Then the owner leaves the room.

Remaining in the room are the dog, the treat, and a quietly observing video camera. Here's the dog's chance to do the wrong thing. What the dogs do is only the beginning of the data for our experiment. In most cases we assume that if given the opportunity, the dog's first move is to get the treat. We wait until he does. Then the owner returns. Here is the crucial data: How does the dog behave?

Every psychological and biological experiment is designed to control one or more variables, while leaving the rest of the world unchanged. A variable can be anything: ingestion of a drug, exposure to a sound, presentation with a set of words. The idea is simply that if this variable is important, the subject's

---

*The command varies from owner to owner—from *no!* to the recently popular *leave it!* Each is fundamentally a negation: a sharp-sounding grammatical flourish that can be applied concurrently to any behavior to make it off-limits.

behavior will be changed when exposed to it. In my experiment, there are two variables: whether the dog eats the treat (the one owners are most interested in) and whether the owner knows whether the dog has eaten it (the one I guessed the dogs are most interested in). Over a handful of trials, I alternate these variables one at a time. First the opportunity to eat the treat is varied: either removing the treat after the owner leaves, providing the dog the treat, or letting the dog stew over it (and eventually disobey). What we tell the owner of the dog's behavior is also varied: in one trial the dog eats the treat, and the owner is informed on return to the room; in another, the dog is surreptitiously given the treat by the videographer, and the owner is misled into thinking that the dog has obeyed the command not to eat.

All the dogs survive the experiment looking well fed and a little bewildered. In many of the trials, the dogs could be models for the guilty look: they lower their gaze, press their ears back, slump their body, and shyly avert their head. Numerous tails beat a rapid rhythm low between their legs. Some raise a paw in appeasement or flick their tongue out nervously. But these guilt-related behaviors did not occur more often in the trials when the dogs had disobeyed than in those when they had obeyed. Instead there were more guilty looks in the trials when the owner scolded the dog, whether the dog had disobeyed or not. Being scolded despite resisting the disallowed treat led to an extra-guilty look.

This indicates that the dog has associated the owner, not the act, with an imminent reprimand. What's happening here? The dog is anticipating punishment around certain objects or when seeing the subtle cues from the owner that indicate he may be angry. As we know, dogs readily learn to notice associations between events. If the appearance of food follows the opening of the large cold box in the kitchen, why, the dog will be alert to the opening of that box. These associations can be forged with events of their making as well as those they observe. Much of what is learned is based, deep down, on making associations: whining is followed by attention, so the dog learns to whine for attention; scratching at the trash can causes it to tip and spill its contents, so the dog learns to scratch to get what's inside. And making certain kinds of messes is sometimes followed much later by the presence of the owner, which is itself quickly followed by the reddening of the owner's face, loud verbiage coming out of the owner, and punishment by that reddened loud owner. The key here is that the mere appearance of the owner around what looks like evidence of destruction can be enough to convince the dog that punishment is imminent. The owner's arrival is much more closely linked to punishment than the garbage emptying the dog engaged in hours earlier. And if that's the case, most dogs will assume a submissive posture on seeing their owners—the classic guilty look.

In this case, a claim about the dog's knowledge of his misdeed is importantly off the mark. The dog may not think of the behavior as *bad*. The guilty look is very similar to the look of fear and to submissive behaviors. It is no surprise, then, to find so many dog owners who are frustrated with attempts to punish a dog for bad behavior. What the dog clearly knows is to anticipate punishment when the owner appears wearing a look of displeasure. What the dog does not know is that he is guilty. He just knows to look out for you.

A lack of guilt does not mean dogs do nothing wrong. They not only do plenty of human-defined wrong things, they sometimes seem to flaunt these things: a half-chewed shoe is paraded in front of a busy owner; you are greeted by a dog merrily exhausted from rolling in defecation. The teddy-bear guard dog looked nothing if not proud when photographed surrounded by the teddy-bear remains. Dogs do seem to play with the fact of our knowing and not knowing something—to get attention (which it generally does) and perhaps just for the sake of playing with knowledge. This is not unlike a child testing the limits of his understanding of the physical world by sitting on his high chair, dropping a cup to the floor . . . and again . . . and again: he is seeing what happens. Dogs do this with different states of attention, knowledge, or alertness of their owners. In this way they come to learn more about what we know, which they can then use to their advantage.

In particular, dogs are quite capable of concealing behavior, acting to deflect attention from their true motives. Given what we know about their understanding of mind, it is entirely within their reach to deceive. And given that it is a rudimentary understanding, their deception is not always very good. This too is childlike, as in the two-year-old child who puts his hands over his eyes to "hide" from a parent: partway to hiding, but not quite getting the essence of "hidden." Dogs show both imaginative insights and inadequacies. They do not work to hide the spoils of an overturned trash can or a messy roll in the grass. But they do act in ways to conceal their true intent. To stretch forward idly next to a dog playing with a treasured toy—only to get close enough to snatch it. To shriek overly dramatically when bitten in play, thereby ending a momentary disadvantage as the playmate stops in shock. These behaviors may begin fortuitously, with accidental actions that turn out to yield happy consequences. Once noticed, they will be produced again and again. It only remains now for an experimenter to provide an opportunity for dogs to intentionally deceive one another—unless they are too clever to let their scheming be revealed.

A dog's age (About emergencies and death)

With age she uses her eyes less; she looks at me less.

With age she would rather stand than walk, lie than stand—and so she lies next to me outside with her head between her legs, nose still alert to the smells on the breeze.

With age she has become more stubborn, insisting on hoisting herself up stairs without help.

With age the difference is amplified between her day mood—reluctant to walk, extra-sniffy—and her evening mood—pulling me out the door, a spring in her step, willing to forsake smells for a jaunty tour around the block.

With age I have been given a gift: the details of Pump's existence have become even more alive. I started seeing the geography of smells she checks up on in the neighborhood; I feel how long are the periods she waits for me; I hear the way she speaks volumes by simply standing; I see her efforts to cooperate when I goad her to trot across the street.

Every dog that you name and bring home will also die. This inescapable, dreadful fact is part of our lot for introducing dogs into our lives. What is less certain is whether our dogs themselves have any inkling of their own mortality. I inspect Pump for any sign that she notices the age of her sniffmates on the sidewalks; notes the disappearance of the old droopy-eared fella with the cloudy eyes from down the block; observes her own slowed and stiff gait, graying fur, and lethargic mood.

It is our grasp of the fragility of our own existence that makes us wary of risky undertakings, cautious for ourselves and those we love. Our mortal knowledge may not be visible in all of our moves, but it shines through in some: we shrink back from the balcony's edge, from the animal with unknown intent; we buckle up for safety; we look both ways before crossing; we don't jump in the tiger cage; we refrain from the third serving of fried ice cream; we even entertain not swimming after eating. If dogs know about death, it might show in how they act.

I would prefer that dogs not know. On the one hand, when I have been confronted with a dying dog, I wanted to be able to explain to her her situation—as though an explanation would be a comfort. On the other, despite many owners' habit of giving explanations to their dogs for every command or event (*come ON,* I overhear regularly in the park, *we've got to go home so Mommy can get to work* . . .), dogs do not seem comforted by explanations. A life untrammeled by knowledge of its end is an enviable life.

There are a few indications that we should not envy them much. One comes from their own balcony aversion: for the most part dogs reflexively withdraw from true danger, be it a high ledge, a rushing river, or an animal with a predatory gleam in its eye. They act to avoid death.

But so does the lowly paramecium, beating a hasty retreat from predators and toxic substances. Avoidance behavior is instinctual, seen in some form in nearly all organisms. Instincts, from the knee jerk to an eye blink, do not require that the animal understands what it is doing. And we are not ready to grant the paramecium an understanding of death. But that reflex is not trivial: a more sophisticated understanding could be bootstrapped onto it.

And here are two ways dogs differ from the paramecium: First, they are not only avoidant of injury, they act differently once injured. They are aware of when they are damaged. Hurt or dying, dogs often make great efforts to move away from their families, canine or human, to settle down and perhaps die someplace safe.

Second, they are attentive to the dangers that others put themselves in. One need not wait long for a story of a heroic dog to pop up in the local news. A child lost in the mountains is kept alive by the warmth of dogs who stayed with him; a man who falls through the ice of a frozen lake is saved by the dog who came to him at the ice's edge; a dog's barking attracts a boy's parents before he can reach into the hole of a poisonous snake. Heroic dogs tales abound. My friend and colleague Marc Bekoff, a biologist who has studied animals for forty years, writes of a blind Labrador retriever named Norman who was roused to action by the screams of the family's children, caught in the current of a raging river: "Joey had managed to reach the shore, but his sister was struggling, making no headway, and in great distress. Norman jumped straight in and swam after Lisa. When he reached her, she grabbed his tail, and together they headed for safety."

The end result of all the dogs' actions is clear: someone was able to avert death for another day. Given that the dogs needed to overcome their own instinct of self-preservation to preserve another self, the usual interpretation is that the dogs are heroic, not inadvertent, actors. An understanding of the dire straits faced by the various humans might seem the only explanation.

But the trouble with anecdotes is that one does not have the full story of what happened, since the teller, with his own umwelt and particular perception, is necessarily restricted in what he sees. One could reasonably ask whether Norman did not as much intend to save Lisa as, say, follow her brother's instruction to swim out to her; or maybe Lisa herself was able to swim to shore on seeing her faithful companion near; or maybe the current shifted and carried her to shore. There is no videotape to rewind and examine to carefully consider what happened here—or in any of the rescues described. Nor do we know the long-term behavior of the dogs. It is one thing if a dog suddenly barks in order to alert others that a boy is imperiled; it is another if that dog is barking all the time, day and night. An understanding of the dogs' life histories is also important to correctly interpret what happened.

Finally, what of all the cases when a dog *didn't* save the drowning child or the lost hiker. The newspaper headlines never crow, LOST WOMAN DIES AFTER DOG FAILS TO FIND AND DRAG HER TO SAFETY! If the heroic dogs are taken to represent the species, so should the non-heroes be given consideration. There are certainly more unreported non-heroic acts than there is reported heroism.

Both the skeptical and the heroic talk can be displaced by a more powerful explanation, wrought by looking more closely at the dogs' behavior. Scrutiny of these dog stories reveals a recurring element: the dog *came toward* his owner, or *stayed close* to the person in distress. The warmth of a dog saves a lost, cold child; a man in a frozen lake can grab on to his dog waiting on the ice. In some cases the dog also created a ruckus: barking, running around, calling attention to himself—and to, say, the venomous snake.

These elements—proximity to the owner, and attention-getting behavior—are by now familiar to us as characteristic of dogs, and go into their being such fine companions for humans.

And in these cases, they were also essential for the survival of the person whose life was at risk. So are the dogs truly heroes? They are. But did they know what they were doing? There is no evidence that they did. And they don't know they're acting heroically. Dogs certainly have the potential, with training, to be rescuers. Even the untrained dog may come to your aid—but without knowing exactly what to do. Their success is due instead to what they *do* know: that something has happened to you, which makes them anxious. If they express that anxiety in a way that attracts other people—people with an understanding of emergencies—to the scene, or allows you leverage out of a hole in the ice, great.

This conclusion is affirmed by one clever experiment performed by psychologists interested in whether dogs show appropriate behavior when there is an emergency. In this test, owners conspired with the researchers to feign emergencies in the presence of their dogs, in order to see how their dogs responded. In one scenario, owners were trained to fake a heart attack, complete with gasping, a clutch of the chest, and a dramatic collapse. In the second scenario, owners yelped as a bookcase (made of particleboard) descended on them and seemed to pin them to the ground. In both cases, the owners' dogs were present, and the dogs had been introduced to a bystander nearby—perhaps a good person to inform if there has been an emergency.

In these contrived setups, the dogs acted with interest and devotion, but not as though there were an emergency. Dogs frequently approached their owners, and sometimes pawed or nuzzled these seeming victims, now silent and unresponsive (in the heart attack case) or crying out for help (in the bookcase scenario). Other dogs, though, took the opportunity to roam around in the vicinity, wandering and sniffing the grass or the floor of the room. In only a very few cases did a dog vocalize—

which might serve to get someone's attention—or approach the bystander who might be able to help. The only dog who touched the bystander was a toy poodle. The poodle leaped into the bystander's lap and settled down for a nap.

In other words, not a single dog did anything that remotely helped their owners out of their predicaments. The conclusion one has to take from this is that dogs simply do not naturally recognize or react to an emergency situation—one that could lead to danger or death.

A killjoy conclusion? Hardly. If dogs lack the concepts *emergency* and *death* this is not to their discredit. One might as well ask a dog if he understands *bicycles* and *mousetraps* and then censure him for responding with a puzzled tilt of the head. A human child is also naïve to these concepts: an infant has to be screamed at as he zeros in on an open electrical outlet; a two-year-old who saw someone hurt would likely do little but cry. They will be *taught* to understand emergency situations—and then the concept of death. So too are some dogs trained, for instance, to alert a deaf companion to the sound of an emergency device, such as a smoke alarm. The teaching of children is explicit, with some procedural elements—*If you hear this alarm, get Mommy;* the dogs' training is entirely reinforced procedure.

What the dogs seem to know is when an *unusual* situation occurs. They are masters of identifying the usual in the world you share with them. You often act in reliable ways: in your own home, you move from room to room, spending long pauses in armchairs and in front of refrigerators; you talk to them; you talk to other people; you eat, sleep, disappear for long stretches into the bathroom; and so on. The environment is fairly reliable, too: it is neither too hot nor too cold; there is no person in the house apart from the ones who have come in the front door; water is not pooling in the living room; smoke is not drifting in the hallway. From that knowledge of the usual world comes some acknowledgment of the unusual fact of someone's odd behavior when injured, or of the dogs' own inability to act as they customarily can.

More than once Pumpernickel got herself in dire straits (once, trapped on a catwalk heading off a building edge; another time, her leash stuck in the elevator doors as the car began to move). I was amazed at how unfazed she appeared—especially as contrasted with my own alarm. It was never she who got herself out of the fix. I believe that I was more worried about her well-being than she was about mine. Still, much of my well-being hinged on her—not on her knowing how to fix dilemmas, great or small, in my life, but rather on her unremitting cheer and constant companionship.

II

WHAT IT IS LIKE

In our attempt to get inside of a dog, we gather small facts about their sensory capacities and build large inferences upon them. One inference is to the experience of the dog: what it actually *feels like* to be a dog; what his experience of the world is. This assumes, of course, that the world is *like* anything to a dog. Perhaps surprisingly, in philosophical and scientific circles there is a bit of debate about this.

Thirty-five years ago, the philosopher Thomas Nagel began a long-running conversation in science and philosophy about the subjective experience of animals when he asked, "What is it like to be a bat?" He chose for his thought experiment an animal whose almost unimaginable way of seeing had only recently been discovered: echolocation, the process of emitting high-frequency shouts and then listening for the sound being reflected back. How long the sound takes to bounce back, and how it is changed, gives the bat a map of where all the objects are in the local environment. To get a rough sense of what this might be like, imagine lying in a dark room at night and wondering if someone is standing at your doorway. Sure, you could resolve the question by turning on the light. Or, bat-like, you could hurdle a tennis ball at the doorway and see if (a) the ball comes back toward you or flies out of the room, and (b) if a grunt is heard at about the time the ball arrives at the threshold. If you're very good, you might also use (c) how far the ball bounces back, to determine if the person is very tubby (in which case the ball loses most of its speed in his belly) or has washboard abs (which will reflect the ball nicely). Bats use (a) and (c), and in lieu of tennis balls they use sound. And they do it constantly and rapidly, as quickly as we open our eyes and take in the visual scene in front of us.

This, appropriately, boggled Nagel's mind. He thought that the bat's vision, and thus the bat's life, are so wildly odd, so imponderable, that it is impossible to know what it is like to be that bat. He assumed that the bat experiences the world, but he believed that that experience is fundamentally subjective: whatever "it is like," it is that way only to that bat.

The trouble with his conclusion has to do with the imaginative leap that we do make every day. Nagel treated an *inter*species difference as something wholly unlike an *intra*species difference. But we are perfectly happy to talk about "what it is like" to be another human being. I do not know the particulars of another person's experience, but I know enough about the feeling of being human myself that I can draw an analogy from my own experience to someone else's. I can imagine what the world is like to him by extrapolating from my own perception and transplanting it with him at its center. The more information I have about that person—physically, his life history, his behavior—the better my drawn analogy will be.

So can we do this with dogs. The more information we have, the better the drawing will be. To this point, we have physical information (about their nervous systems, their sensory systems), historical knowledge (their evolutionary heritage, their developmental path from birth to adults), and a growing corpus of work about their behavior. In sum, we have a sketch of the dog umwelt. The parcel of scientific facts we have collected allows us to take an informed imaginative leap inside of a dog— to see what it is like to be a dog; what the world is like from a dog's point of view.

We have already seen that it is smelly; that it is well peopled with people. On further consideration, we can add: it is close to the ground; it is lickable. It either fits in the mouth or it doesn't. It is in the moment. It is full of details, fleeting, and fast. It is written all over their faces. It is probably nothing like what it is like to be us.

It is close to the ground . . .

One of the most conspicuous features of the dog is one of the most conspicuously overlooked when contemplating their view of the world: their height. If you think that there is little difference between the world at the height of an average upright human and that at the height of an average upright dog—one to two feet—you are in for a surprise. Even putting aside for a moment the difference in sound and smell close to the ground, being at a different height has profound consequences.

Few dogs are human-height. They are human-knee height. One might even say they are often *underfoot*. We are magnificently obtuse when it comes to imagining even the simple fact of their being less than half our height. Intellectually we know that dogs are not at our height, yet we set up interactions such that the height difference is a constant problem. We put things "out of reach" of dogs, only to be frustrated by their attempts to get them. Even knowing that dogs like greeting us at eye level, we typically do not bend down. Or, bending down just far enough to allow them to reach our faces with a leap, we may get annoyed when they then leap. *Jumping up* is the direct result of desiring to get to something one needs to *jump up* to reach.

Scolded enough for jumping up, dogs happily find there is plenty of interest underfoot. There are, for instance, lots of feet. Smelly feet: the foot is a good source of our signature odors. We tend to sweat pedally when we are mentally taxed: stressed, or concentrating hard. Clumsy feet: sitting, we dangle them, but not with dexterity. They act as single units, with toes only existing as places between which extra odors may be discovered by a roving tongue.

If the foot smells so interesting, of course, then the way we treat them must be awfully frustrating: damned shoes. We cloister our odors. On the other hand, shoes left behind smell just like the person who had been in them, and they have the additional interest of carrying on their soles whatever you squishily stepped in outside. Socks are equally good carriers of our odor, hence the gaping holes that regularly appear in socks left bedside. On examination, each hole has been lovingly poked by the incisors of a dog with a sock in her mouth.

Besides feet, at dog height the world is full of long skirts and trouser legs dancing with every footfall of their wearer. The tight whirling motions the warp of a pant leg presents to a dog's eye must be tantalizing. Between their sensitivity to motion and their investigatory mouths, it is no wonder one can find one's pants being nipped by the dog at the end of your leash.

The world closer to the ground is a more odoriferous one, for smells loiter and fester in the ground, while they distribute and disperse on the air. Sound travels differently along the ground, too: hence birds sing at tree height, while ground dwellers tend to use the earth to communicate mechanically. The vibration of a fan on the floor might perturb a dog nearby; likewise, loud sounds bounce more loudly off the floor into resting dog ears.

The artist Jana Sterbak tried to capture a dog's-eye view by rigging a video camera to a girdle worn by Stanley, her Jack Russell terrier, and recording his perambulations along a frozen river and through Venice, the "city of doges" (pun probably intended). The result is a manic, jumbled rush of sights, the world akilter and the image never calm. At fourteen inches above the ground, Stanley's visual world is a glimpse of his olfactory world: what catches his olfactory interest he pursues in body and sight.

But by suiting up animals with critter-cams we are mostly getting an idea of their *vantage* on the world, not their entire umwelt. With most if not all wild animals, only by taking such a vantage may we have any information about their world, their day: we can't keep up with a diving penguin as a camera strapped to its back can; only an inconspicuous camera could capture the tunnel building of a naked mole rat underground. To watch Stanley from the vantage of his back is to be surprised at the view. There is the temptation, though, to think that by capturing a picture of Stanley's day we have completed the imaginative exercise. It is but the beginning.

... It is lickable ...

She is lying on the ground, head between paws, and notices something potentially interesting or edible a short stretch away on the floor. She pulls her head forward to it, her nose—that beautiful, robust, moist nose—nearly but not quite *in* the particle. I can see her nostrils working to identify it. She gives a wet snort and brings her mouth to aid in the investigation: by turning her head ever-so-slightly on an angle her tongue reaches the floor. She test-licks it with quick swipes, then straightens up and sets to a more serious posture from which to lick, lick, lick the floor—long strokes with the fullness of her tongue.

Nearly everything is lickable. A spot on the floor, a spot on herself; the hand of a person, the knee of a person, the toes of a person, the face, ears, and eyes of a person; a tree trunk, a bookshelf; the car seat, the sheets; the floor, the walls, the all. Unidentifiables on the ground are especially ripe for tonguing. This is revealing, for licking—bringing molecules into oneself, not merely taking a distant safe stance toward them—is an extremely intimate gesture. Not that dogs mean to be intimate. But to be so directly in contact with the world, intentionally or not, is to define oneself differently with respect to one's environment than humans do: it is to find less of a barrier at the edge of one's own skin or fur from that which surrounds it. No wonder it is not unusual to see a dog duck his head fully into a mud puddle or twist his supine body in exaltation of spirit and the rank earth.

The dog's sense of personal space reflects this intimacy with the environment. All animals have a sense of comfortable social distance, the breaching of which causes clashes and the stretching of which they try to contain. While Americans balk at strangers standing closer than eighteen inches, American dogs' personal space is approximately zero to one inches. Repeating itself on sidewalks across the country this very second is a scene that demonstrates the clash of our senses of personal space: the sight of two dog owners as they stand six feet apart, straining to keep their leashed dogs from touching, while the dogs strain mightily to touch each other. Let them touch! They greet strangers by getting into each other's space, not staying out of it. Let them get into each other's fur, sniff deeply, and mouth each other in greeting. It is not for dogs the safe distance of a handshake.

As we have a limit to the proximity of others we'll endure, we also have a limit to the distance we prefer: a kind of social space. Sitting over five or six feet apart makes for an uncomfortable conversation. Walking on opposite sides of the street, we do not feel we are walking *together.* Dogs' social space is more elastic. Some dogs happily walk in parallel but at great, owner-distressing distances from their owners; others like to trot at your heels. This extends to their sense of fit with us, resting at home. Dogs have their own version of enjoying the pleasantness of a book that fits closely but not too tightly into a box. Pump wanted to sit so that her body was cupped by the embrace of a small upholstered chair. She would fill the space created by my bent legs when lying on my side in bed. Other dogs position themselves with the length of their backs against the length of a sleeping body. The pleasure of this alone is enough for me to invite a dog onto the bed.

. . . It either fits in the mouth or it's too big for the mouth . . .

Of the innumerable objects we see around us, only a very few are salient to the dog. The array of furniture, books, tchotchkes,

and miscellany in your home is reduced to a more simple classi-
ficatory scheme. The dog defines the world by the ways that he
can *act* on the world. In this scheme, things are grouped by
how they are manipulated (chewed, eaten, moved, sat upon,
rolled in). A ball, a pen, a teddy bear, and a shoe are equivalent:
all are objects that one can get one's mouth around. Likewise,
some things—brushes, towels, other dogs—act on them.

The affordances—the typical use, the functional tone—that
we see in objects are superseded by dog affordances. A dog is less
threatened by a gun than interested in seeing if it fits in his mouth.
The range of gestures you make toward your dog is reduced to
those that are fearsome, playful, instructive—and those that are
meaningless. To a dog, a man raising his hand to hail a cab says the
same thing as a man reaching to high-five or one waving good-
bye. Rooms have a parallel life in the dog's world, with areas that
quietly collect smells (invisible detritus in the crook of the wall and
floor), fertile areas from which objects and odors come (closets,
windows), and sitting areas where you or your identifying perfume might be found. Outside, they do not so much notice *buildings:* too big; not able to be acted on; not meaningful. But the
building's *corner,* as well as lampposts and fireplugs, wears a new
identity each encounter, with news of other dog passersby.

   For humans it is the form or shape of an item that is usually
its most salient feature, leading to our recognition of it. Dogs, by

contrast, are generally ambivalent about the shape in which, say, their dog biscuits come (it is *we* who think they should be bone-shaped). Instead, motion, so readily detected by the retinae of dogs, is an intrinsic part of the identity of objects. A running squirrel and an idle squirrel may as well be different squirrels; a skateboarding child and a child holding a skateboard are different children. Moving things are more interesting than still ones—as befits an animal at one time designed to chase moving prey. (Dogs will stalk motionless squirrels and birds, of course, once they have learned that they often spontaneously become squirrels running and birds on the wing.) Rolling quickly on a skateboard, a child is exciting, worth barking at; stop the skateboard, and the motion, and the dog calms.

Given their definition of objects by motion, smell, and mouthability, the most straightforward items—your own hand—may not be straightforward to your dog. A hand patting his head is experienced differently than one pressing continuously on it. Similarly, a glance, even many stolen glances, is different than a stare. A single stimulus—a hand, an eye—can become two things when experienced at different speeds or intensities. Even for humans, a series of still images shuffled fast enough becomes a continuous image: as though changing identity. To the common snail, wary of the world, a slowly tapping stick is risky to walk over; but if the stick is oscillated four times a second, the snail will move into it. Some dogs will endure a pat on the head but not a hand resting there; for others, the reverse is true.*

These ways of defining the world can all be seen by watching a dog interact with the world. Dogs entranced by a blank spot on

---

*For a horse, releasing pressure on the body is sufficiently pleasurable as to be able to be used as a reinforcement in training. Perhaps it would be the same with dogs who startle at the feeling of a hand pressed firmly on their head.

the sidewalk, those whose ears perk at "nothing," those transfixed by an invisibility in the bushes—you are watching them experience their sensory parallel universe. With age the dog will "see" more objects familiar to us, will realize that more things can be mouthed, licked, rubbed against, or rolled in. They also grow to understand that different-seeming objects—the man at the deli, and the deli man on the street—are one and the same. But whatever we think we see, whatever we think just happened in a moment, we are pretty much assured that dogs see and think something different.

... It is full of details ...

Part of normal human development is the refinement of sensory sensitivity: specifically, learning to notice less than we are able to. The world is awash in details of color, form, space, sound, texture, smell, but we can't function if we perceive everything at once. So our sensory systems, concerned for our survival, organize to heighten attention to those things that are essential to our existence. The rest of the details are trifles to us, smoothed over, or missed altogether.

But the world still holds those details. The dog senses the world at a different granularity. The dog's sensory ability is sufficiently different to allow him to attend to the parts of the visual world we gloss over; to the elements of a scent we cannot detect; to sounds we have dismissed as irrelevant. Neither does he see or hear everything, but what he notices includes what we do not. With less ability to see a wide range of colors, for instance, dogs have a much greater sensitivity to contrasts in brightness. We might observe this in their reluctance to step into a reflective pool of water, in a fear of entering a dark room.* Their sensitivity to motion alerts them to the deflating balloon wafting gently curbside. Without speech, they are more attuned to the prosody in our sentences, to tension in our voice, to the exuberance of an exclamation point and the vehemence of capital letters. They are alert to sudden contrasts in speaking: a yell, a single word, even a protracted silence.

As with us, the dog's sensory system is attuned to novelty. Our attention focuses on a new odor, a novel sound; dogs, with a wider range of things they smell and hear, can seem to be constantly at attention. The wide-eyed look of a dog trotting down the street is that of someone being bombarded with the new. And, unlike most of us, they are not immediately habituated to the sounds of human culture. As a result, a city can be a explosion of small details writ large in the dog's mind: a cacophony of the everyday that we have learned to ignore. We know what a car door slamming sounds like, and unless listening for just that sound, city dwellers tend to not even hear the symphony of slams playing on the street. For a dog, though, it may be a new sound each time it happens—and one that sometimes, even more interestingly, is followed by a person arriving on the scene.

They pay attention to the slivers of time between our blinks, the complement of what we see. Sometimes these are not invisible things but simply those we would prefer they not pay attention to, like our groins, or the favored squeaking toy we stuff in a pocket, or the forlorn, limping man on the street. We could see those things, too, but we look away. Human habits that we ignore—tapping our fingers, cracking our ankles, coughing

*As Temple Grandin has similarly noted with cows and pigs, causing the meat industry to alter the paths the animals walk into the slaughterhouse. For the industry, her work is useful in promoting less stressed, and thus better-tasting meat. For the animals, they are presumably spared from some added anxiety as they travel— one hopes unknowingly—toward their deaths.

politely, shifting our weight—dogs notice. A shuffle in a seat—
it may foretell rising! A scootch forward in the chair—surely
something is happening! Scratching an itch, shaking your head:
the mundane is electric—an unknown signal and a whiff of
shampoo. These gestures are not part of a cultural world for dogs
as they are for us. Details become more meaningful when they
are not swallowed up in the concerns of the everyday.

That very attention that dogs bring to us may cause them to
acclimate to these sounds over time, to be inculcated in the
human culture. Watch a bookstore dog, who lives out the hours
of his day surrounded by people: he has become inured to
strangers coming by, standing close while they riffle the pages of
a book; to being scratched on the head, to passing smells and
ever-present footsteps. Crack your knuckles a dozen times a
day and a nearby dog will learn to ignore this habit. By contrast,
a dog unaccustomed to human habits is alarmed at every one:
the most exciting and frightening thing that could happen to a
dog left chained to guard a house is that it actually requires his
guarding. Guard dogs may only occasionally see an unknown
person walking by, a new smell on the air or new sound, let
alone any rampant knuckle-crackers.

We can begin to make up for our human disadvantage in
understanding the dog's sensory umwelt by trying to startle our
sensory systems. For instance, to escape our bad habits of seeing
things roughly in the same colors every day, expose yourself to a
room lit by only one color—say a narrow bandwidth of yellow.
The colors of objects under such light are washed out: your
own hands are drained of their blood-filled vitality; pink dresses
turn dully white; face stubble stands out like pepper in a bowl of
milk. The familiar is made foreign. But for the yellow glow from
above, this is much closer to what it might be like to have a dog's
color perception.

. . . It is in the moment . . .

Ironically, attention to details may preclude an ability to generalize from the details. Sniffing the trees, the dog does not see the forest. Specificity of place and object is useful when you want to calm your dog on a road trip: you can bring his favored pillow to help calm him. A feared object or person put in a new context can sometimes be reborn as unscary.

That same specificity might indicate that dogs do not think abstractly—about that which is not directly in front of them. The influential analytic philosopher Ludwig Wittgenstein suggested that though a dog can *believe* that you are on the other side of the door, we cannot sensibly talk of his *ruminating* on it: believing that you will be there in two days' time. Well, let's eavesdrop on that dog. He has slowly zigzagged through the house since you left. He has run through all the interesting unchewed surfaces in the room. He has visited the armchair, where food was once left unattended long ago, and the couch, where food was spilled last night. He has napped six times, had three visits to the water bowl, lifted his head twice at faraway barks. Now he hears your shuffling approach of the door, quickly confirms by nose that it is you, and remembers that each time he hears and smells you, you appear visually next.

In sum, he believes that you are there. It is nonsense to suggest otherwise. Wittgenstein's doubt is not that dogs have beliefs. They have preferences, make judgments, distinguish, decide, refrain: they think. Wittgenstein's doubt is that before you arrive, your dog is anticipating your arrival: pondering it. It is doubt that dogs have beliefs about things not happening right now.

To live without the abstract is to be consumed by the local: facing each event and object as singular. It is roughly what it means to live *in the moment*—to live life unburdened by reflection. If it is so, then it would be fair to say that dogs are not reflective. Though they experience the world, they are not also considering their own experiences. While thinking, they are not consulting their own thoughts: thinking about thinking.

Dogs come to learn the cadence of a day. But the nature of a moment—the experience of moments—is different when olfaction is your primary sense. What feels like a moment to us may be a series of moments to an animal with a different sensory world. Even our "moments" are briefer than seconds; they are the duration of a noticeable instant, perhaps the smallest distinguishable time unit, as we normally experience the world. Some suggest that this is measurable: it is an eighteenth of a second, the length of time a visual stimulus has to be presented to us before we consciously acknowledge it. Thus we barely notice a blink of an eye, at a tenth of a second long. By this logic, with a higher flicker-fusion rate, a visual moment is briefer and quicker for dogs. In dog time each moment lasts less long, or, to put it another way, the next moment happens sooner. For dogs, "right now" happens before we know it.

. . . It is fleeting and fast . . .

For dogs, perspective, scale, and distance are, after a fashion, *in* olfaction—but olfaction is fleeting: it exists in a different time scale. Scents don't arrive with the same even regularity as (under normal conditions) light does to our eyes. This means that in their scent-vision they are seeing things at a different rate than we.

Smell tells time. The past is represented by smells that have weakened, or deteriorated, or been covered. Odors are less strong over time, so strength indicates newness; weakness, age.

The future is smelled on the breeze that brings air from the place you're headed. By contrast, we visual creatures seem to look mostly in the present. The dogs' olfactory window of what is "present" is larger than our visual one, including not just the scene currently happening, but also a snatch of the just-happened and the up-ahead. The present has a shadow of the past and a ring of the future in it.

In this way, olfaction is also a manipulator of time, for time is changed when represented by a succession of odors. Smells have a lifetime: they move and they expire. For a dog, the world is in flux: it waves and shimmers in front of his nose. And he must keep sniffing—as if we had to repeatedly look at and attend to the world for a constant image to remain on our retinae and in our minds—for the world to be continually apparent to him. This explains so much familiar behavior: your dog's constant sniffing, for one,* and also, perhaps, his seemingly divided attention, which races from sniff to sniff: objects only continue to exist as long as an odor is emitted and he inhales. While we can stand in one place and take in a view of the world, dogs must do much more moving themselves in order to absorb it all. No wonder they seem distracted: their present is constantly moving.

The odor of objects thus holds the data of passing minutes and hours. As they note the hours and days, dogs can note the seasons through smell. We on occasion notice the passing of a season as marked by the smell of blooming flowers, decaying leaves, air about to burst into rain. Mostly, though, we feel or see the seasons: we feel the welcome sun on our winter-paled skin; we glance out the window on a bright spring day and never remark, *What a beautiful new smell!* Dogs' noses stand in for our

---

*To pull a dog from ardent sniffing is the same for him as being yanked away from a scene just as soon as you turn your eyes to it.

sight and skin sense. The air of spring brings odors in every sniff-ful remarkably different from the air of winter: in its moisture or heat; the amount of rotting death or blooming life; in air traveling on breezes or emanating from the earth.

Navigating the world of human time with their expanded window of the present, dogs function a little ahead of us; they are preternaturally sensitive, a shade faster. This accounts for their skill at catching the tossed ball midair and also for some of the ways they seem out of sync with us, some of the ways we can't get them to do what we want. When dogs don't "obey," or have difficulty learning something we want them to, it is often that we are not reading *them* well: we don't see when their behavior has begun.* They are lunging toward the future a step before us.

... It is written all over their faces ...

She has a smile. It's one of the panting faces she puts on. Not every panting face is a smile, but every smile is a panting face. A slight fold in her lip—it would be a dimple on a human face—adds to the smile. Her eyes can be saucers (engaged) or half-open slits (contented). And her eyebrows and eyelashes exclaim.

Dogs are ingenuous. Their bodies do not deceive, even if they sometimes cajole or trick us. Instead the dog's body seems to

*Clicker training tries to address this dissonance of our different "moments" and our different senses of what the dog is "doing" at any moment. Trainers use a small device that allows them to make a sharp, distinct *click!* when the dog has done a desired behavior and can expect an imminent reward. The click helps make a human moment salient to a dog; left to his own devices, the dog parcels up his life differently.

map straight to his internal state. Their joy when you return home or when you approach them is translated directly through their tails. Their concern is plotted by the lift of an eyebrow. Pump's smile is not an actual grin, but that deep lip retraction that gives a glimpse of teeth *is* used in a ritualized way, part of a communication with us.

You can tell a lot about a dog by observing how he carries his head. Mood, interest, and attention are writ in capital letters from the altitude of the head, the lay of the ears, and the radiance of the eyes. Think of a dog prancing around in front of other dogs, tail and head high, with a cherished or stolen toy: given dogs' usual way of negotiating around each other, this is a clear, intentional gesture—of something like pride. Young wolves too may cheekily flaunt food in front of older animals. The leader in interaction with the world, the head is usually aimed in the direction the dog is going. If a dog turns his head to the side, it is just momentary—to determine if there is something worth pursuing yonder. This is unlike us, who might turn our heads in contemplation, to strike a pose, or for effect. The dog is refreshingly free of pretense.

What the head doesn't tell of the dog's intent, the tail does. The head and tail are mirrors, conveying the same information in parallel media, the classic antithesis. But they can also be true pushmi-pullyus, differently sensitive at either end. A dog who balks at being sniffed in the face may be fine being examined at the rump, or vice versa. Either the tail or head is telling you what is inside.

I would be more surprised if I were entirely correct about "what it is like" inside of a dog than if I am entirely wrong. To address this question is to begin an exercise in empathy, informed imagination, and perspective-taking more than it is to discover the conclusive account. Nagel suggested that no objective account can ever be made of other species' experiences. The privacy of the dog's personal thoughts is intact. But it is crucial that we try to imagine how he sees the world—that we replace anthropomorphisms with umwelt. And if we look carefully enough, imagine skillfully enough, we may surprise our dogs with how much we get right.

You Had Me at Hello

I walk in the door and waken Pump with my arrival. First, I hear her: the thump-thumping of her tail against the floor; her toenails scratching on the ground as she rises, heavily; the jingle of collar tags as she wriggles a shake down the length of her body and out her tail. Then I see her: her ears press back, her eyes soften; she smiles without smiling. She trots to me, her head slightly down, ears perked and tail swinging. As I reach forward she snuffles a greeting; I snuffle back. Her moist nose just touches me, her whiskers sweep my face. I'm home.

Here's a possible explanation for why dogs were not the subjects of serious scientific inquiry until recently: you don't ask questions when you already know the answers viscerally. The delight of my twice- or thrice-daily reunions with Pumpernickel is matched by their ordinariness. Nothing could seem more natural than these simple interactions: they are wonderful, but it is not a wonder that at once demands scientific scrutiny. I may as well dwell on the nature of my right elbow: it is simply a part of me, all the time, and I don't puzzle over its helpful placement there precisely between my upper arm and my forearm, or ponder what it might be like in the future.

Well, I should reconsider that elbow. For the nature of what in certain circles is termed the "dog-human bond" is exceptional. It is not just any animal awaiting my arrival, and it is not just any dog. It is a very particular kind of animal—a domesticated one— and a particular kind of dog—one with whom I have created a symbiotic relationship. Our interactions enact a dance to which only we know the particular steps. Two things—domestication and development—made the dance possible at all. Domestication sets the stage; the rituals are created together. We are bound together before we know it: it is before reflection or analysis.

The human bond with dogs is animal at its core: animal life has succeeded by individual animals associating with, and eventually bonding with others. Originally animals' connection with each other may only have lasted for one sex-filled instant. But the meeting of anatomy at some point evolved in myriad directions: into long-term pairings centered around raising young; groups of related individuals living together; unions of same-sex, non-mating animals for protection or companionship or both; even alliances between cooperative neighbors. The classic "pair bond" is a description of the association that forms between two mated animals. Bonded animals might be recognized by even a naïve observer: most pair bonds hang out together. They mind and care for each other, and they excitedly greet each other on reuniting.

This kind of behavior may seem unsurprising. After all,

we humans spend much time trying to pair-bond, maintaining or discussing our current pair bonds, or trying to extricate ourselves from ill-advised pair bonds gone sour. But from an evolutionary point of view, bonding with others is non-obvious. The goal of our genes is to reproduce themselves: an inherently selfish aim, as sociobiologists observe. Why bother with others at all? The explanation of a selfish gene bothering to mind and greet other gene forms turns out to also be selfish: sexual reproduction increases the chance of helpful mutations. It also behooves the selfish gene to ensure that one's sexual mate is healthy enough to bear and raise the new, infant genes.

Sound far-fetched? A biological mechanism has been discovered that supports pair-bonding. Two hormones, oxytocin and vasopressin (known for their roles in, respectively, reproduction and body-water regulation), are released when interacting with one's partner. These hormones make changes at the neuronal level, in areas of the brain involved in pleasure and reward. The neural change results in a behavioral change: encouraging association with one's mate, because it simply feels good. In the small, mouselike prairie voles that the researchers studied, the vasopressin seems to work on dopamine systems, which results in the male vole being very solicitous of his mate. As a result, prairie voles are monogamous, forming long-lasting pair bonds, in which both parents are involved in raising the wee voles.

But these are intraspecific pair bonds: between members of the same species. What started cross-species bonding, which now results in our living with, sleeping with, and dressing up in sweaters our dogs? Konrad Lorenz was the first to describe it. He gave a description of what he called simply "the bond" in the 1960s, well before the current age of neural science, and before human-pet relationship seminars. In scientific language, he defined the bond as revealed in "behaviour patterns of an objectively demonstrable mutual attachment." In other words, he redefined the bond between animals not by its goal—such as mating—but by the process—such as co-habitating and greeting. The goal could be to mate, but it could also be survival, work, empathy, or pleasure.

This refocus opens the door to considering lots of other, non-mating kinds of pairings as true bonds—between members of the same species, or between two species. Among dogs, working dogs are a classic case. For instance, sheepdogs bond early in life with the intended subject of their work: sheep. In fact, to be effective herders, sheepdogs must bond with sheep in their first few months. They live among the sheep, eat when the sheep eat, and sleep where the sheep sleep. Their brains are in the throes of rapid development at an early age; if they don't meet sheep then, they don't become good shepherds. All wolves and dogs, working or not, have sensitive periods of social development. Early in puppyhood they show a preference for the caregiver, seeking her out and responding to her differently than to others, with a special greeting.* For young animals, it is adaptive to do so.

There's still a big leap, though, between a bond wrought of developmental advantage and one based in companionship. Given that humans neither mate with dogs nor need them to survive, why might we bond?

---

*This might seem a good time for a young dog to meet his new owner. There is surprisingly little good science about the timing of this introduction. The forces determining when people adopt dogs are more often than not influenced by everything *but* the best age for a puppy to meet a person. Many states have laws prohibiting sale of puppies prior to 8 weeks, to protect against selling physically immature animals. Breeders have their own interests in mind in selling their charges. But social recognition requires experience. From two weeks to four months dogs are particularly open to learning about others (of any species). No dog should be taken away from his mother before he is weaned (which can be from six to ten weeks), but dogs *should* be exposed to humans as well as to littermates.

BONDABLES

The feeling of mutual responsiveness: that each time one of us approached or looked at the other, it *changed* us—it effected some response. I smiled to see her look or wander over; her tail would thump and I could see the slight muscle movements of the ears and eyes that indicated attention and pleasure.

We don't need to be herded; neither are we born to herd. Nor, as we saw earlier, are we a natural pack. What, then, accounts for our bond with dogs? There are a number of characteristics of dogs that make them good candidates for us to choose to bond with. Dogs are diurnal, ready to be awake when we can take them out and asleep when we can't. Notably, the nocturnal aardvark and badger are rare as pets. Dogs are a good size, with enough variation between breeds to suit different specs: small enough to pick up, big enough to take seriously as an individual. Their body is familiar, with parts that match ours— eyes, belly, legs—and an easy mapping on most of those that don't—their forelimbs to our arms; their mouth or nose to our hands.* (The tail is a disparity, but it is pleasing in its own right.) They move more or less the way we do (if more swiftly): they go forward better than backward; they have a relaxation to

---

*We are generally enthralled by creatures that look like us in at least some way. Notably, not every and not all animals are effused over, taken in, or anthropomorphized: monkeys and dogs regularly are, but eels and manta rays rarely are. "That barnacle just loves hanging out with me and my boat" is a sentence never uttered. The difference between the monkey and the barnacle is part evolutionary, part familiarity. An infant monkey curling a hand around a mother's finger easily evokes the same poignant scene between human mothers and infants. By contrast, however much a young eel may be yearning for contact as it slides toward its mother, its lack of limbs gets in the way of our calling the scene "touching"—or even intentional.

their stride and a grace to their run. They are manageable: we can leave them by themselves for long stretches of time; their feeding is not complicated; they are trainable. They try to read us, and they are readable (even if we often misread). They are resilient and they are reliable. And their lifetime is in scale with ours: they will oversee a long arc of our lives, perhaps from childhood to young adulthood. A pet rat might live a year—too brief; the gray parrot sixty—too long; dogs hit a middle ground.

Finally, they are compellingly cute. And by compelling, I mean a literal compulsion: it is part of our constitution that we coo over puppies, that we soften at the sight of a big-headed, small-limbed mutt, that we go ga-ga for a pug nose and a furry tail. It has been suggested that humans are adapted to be attracted to creatures with exaggerated features—the prime examples of which are human infants. Infants come with comically distorted versions of adult parts: enormous heads; pudgy, foreshortened limbs; teeny fingers and toes. We presumably evolved to feel an instinctual interest in, and drive to help, infants: without an older human's assistance, no infant would survive on its own. They are adorably helpless. Thus those non-human animals with neotenized (infantlike) features may prompt our attention and care because these are features of human juveniles. Dogs accidentally fit the bill. Their cuteness is half fur and half neoteny, which they have in spades: heads overly large for their bodies; ears all out of proportion with the size of the heads they are attached to; full, saucer eyes; noses undersized or oversized, never nose-sized.

All these features are relevant in attracting us to dogs, but they don't fully explain why we bond. The bond is formed over time—not just on looks, but on how we interact together. At its most general, the explanation may simply be that, as one of Woody Allen's characters says, we need the eggs. He describes his own crazy pair-bonding attempts with a joke about his brother, a fellow so off that he thinks he's a chicken. Sure, the family could send him to be fixed of this delusion, but they're too happy with the protein-rich spoils of his mental disease. In other words, the answer is a non-answer: it's simply in our nature to bond.* Dogs, who evolved among us, are the same way.

At a more scientific level, the question of how bonding came to be in the nature of dogs and humans is answerable in two ways: with explanations that in ethology are called "proximate" and "ultimate." An ultimate explanation is an evolutionary one: why a behavior like bonding to others evolved to begin with. The best answer here is that both we and dogs (and dogs' fore-bears) are social animals, and we are social because it turned out to confer an advantage. For instance, one popular theory is that human sociality allowed for the distribution of roles that enabled them to hunt more effectively. Thus our ancestors' success at hunting made it possible for them to survive and thrive, while those poor Neanderthals who stuck it out on their own did not. For wolves, too, staying in social family groups allows for coop-erative hunting of large game, for the convenience of a mating partner, and for assistance in rearing the pups.

We might be social with any other social animal; but we do not, notably, bond with meerkats, ants, or beavers. To explain our particular choice of dogs, we must look one step more immediate. A proximate explanation is a local one: what imme-diate effect the behavior has that reinforces it, or rewards the

*Edward O. Wilson, the naturalist and sociobiologist who studied ant popula-tions in amazing detail, proposed that we have an inborn, species-typical tendency to affiliate with other animals: what has been called the "biophilia hypothesis." The notion is attractive and also much debated. It is, notably, difficult to disprove such a hypothesis. Regardless, I consider it the scientist's way of saying what Woody Allen did.

"behaver." For an animal, reinforcement could be the meal that follows a hunt or the copulation that follows an ardorous, energetic pursuit.

It is here that dogs distinguish themselves from the other social animals. There are three essential behavioral means by which we maintain, and feel rewarded by, bonding with dogs. The first is contact: the touch of an animal goes far beyond the mere stimulation of nerves in the skin. The second is a greeting ritual: this celebration of encountering one another serves as recognition and acknowledgment. The third is timing: the pace of our interactions with each other is part of what can make them succeed or fail. Together, they combine to bond us irrevocably.

TOUCHING ANIMALS

Neither of us is truly comfortable but neither of us moves. He is on my lap, sprawled across my thighs, his legs already a little long and dangling down the side of the chair. He's settled his chin on my right arm, right in the crook of my elbow, his head tilted sharply upward just to keep in contact with me. To type, I must strain to pull my trapped arm up and just over the desktop onto the keyboard, with only my fingers able to move freely, and my body leaning precariously. We're both working to hold on to each other, to keep that gossamer of contact that says we are going to intertwine our fates—or they are already intertwined.

We named him Finnegan. We found him at a local shelter, in a cage among dozens of cages, in a room among a dozen rooms, all filled with dogs who we could just as easily have taken home. I remember the moment I knew it would be Finnegan. He leaned. Outside of his cage, on the tabletop where germ-carrying humans were allowed to interact with the sick dogs, he wagged, his ears flopped around his tiny face, he coughed long bursts of coughs, and he leaned against my chest, at table height, his face tucked into my armpit. Well, that was that.

Often it is contact that draws us to animals. Our sense of touch is mechanical, matter on matter: different than our other sensory abilities, and arguably more subjectively determined. The stimulation of a free nerve ending in the skin could be, depending on the context and the force of stimulation, a tickle, a caress, unendurable, painful, or unnoticed. If we are distracted, what would otherwise feel like a painful burn might be a niggling irritation. A caress might be a grope if it comes from an unwanted hand.

In our current context, though, "touch" or "contact" is simply the erasing of a gap separating bodies. Petting zoos have arisen to satisfy the urge to engage that animal on the other side of the fence not only by looking at it, but by *touching* it. Better still if the animal is touching back—with, say, a warm tongue or worn teeth grabbing at the food in your outstretched hands. Children and even adults who approach me on the street as I walk with my dog want not to look at the dog, to watch her wag, to meditate on the dog—no, they want to pet the dog: to touch her. In fact, after a cursory rub, many people appear satisfied with that interaction. Even a brief touch is sufficient to bolster the feeling that a connection has been made.

> Occasionally one might find one's toes, hanging off the end of the bed bare, being licked.

Dogs and humans share this innate drive for contact. The contact between mother and child is natural: by dint of the requirement for food, the infant is drawn to the mother's breast.

Thenceforth, being held by the mother may be naturally comforting. A child who has no caregiver, male or female, will develop abnormally, in ways that it would be inhuman to experimentally test. Inhumane or not, in the 1950s a psychologist named Harry Harlow enacted a series of now notorious experiments designed to test the importance of maternal contact. He took infant rhesus monkeys away from their mothers and raised them in isolation. Some had the choice of two surrogate "mothers" in their enclosures: a wire-framed, monkey-sized doll covered in cloth, plumped with filling, and warmed with a lightbulb; or a bare wire monkey with a bottle full of milk. Harlow's first discovery was that the infant monkeys spent nearly all their time huddled against the cloth mother, dashing over to the wire mother periodically to feed. When exposed to fearful objects (demonic noise-making robotic contraptions Harlow put in their cages), the monkeys tore for the cloth mothers. They were desperate for contact with a warm body—with just that warm body from which they had been removed.*

The long-term discovery from Harlow's work was that these isolated monkeys developed relatively normally physically, but abnormally socially. They did not interact with other monkeys well: terrified, they huddled in the corner when another young monkey was put into their cage. Social interaction and personal contact is more than desirable: it is necessary for normal development. Months later, Harlow tried to rehabilitate those monkeys whose early isolation so malformed them.

---

*In studies with puppies, researchers found that those distressed at separation from their mothers and littermates vocalized somewhat less if given a towel or soft toy (a stuffed blue lamb). If there is knowledge to be gained here it is that a soft familiar object can be a salve (hence, in children, the power of teddy bears); in fact, such an object may reduce some of the unease dogs may manifest at being left at home alone.

He found that the best remedy was regular contact with young normal monkeys, whom he came to call "therapy monkeys," in play. This restored some of the isolates to more normal social actors.

Watch an infant child, with limited vision and even more limited mobility, try to snuggle into his mother, his head rooting around for contact, and one is seeing just what newborn puppies look like. Blind and deaf at birth, they are born with the instinct to huddle with siblings and their mother, or even with any solid object nearby. The ethologist Michael Fox describes the head of a puppy as a "thermotactile sensory probe," moving in a semicircle until it touches something. This begins a life of social behavior reinforced by and embracing contact. Wolves are estimated to make a move to touch one another at least six times an hour. They lick—each other's fur, genitals, mouths, and wounds. Snouts touch snouts or body or tail; they nuzzle muzzles or fur. They are oriented to touch even in agonistic activity, which, unlike many other species, usually involves contact: pushing, pinning with a bite, biting the body or leg, seizing another's muzzle or head with one's mouth.

Directed toward us, the dog's youthful instinct becomes a drive to burrow a head under our sleeping bodies or to rest a head upon us; to push and bump us as we walk; to gently nibble or lick us dry. It seems no accident that dogs playing at full steam regularly run into any observing owners nearby, using them as living bumpers defining their playing field. In turn, dogs suffer being touched by us. This is to their infinite credit. We find them touchable: furry and soft, right under dangling fingertips and often wearing their neoteny to greatly cute result. The dog's experience of that touch, though, is likely not what we think. A child may rub the belly of a dog fiercely; we reach to pat a dog's head—unknowing whether they want to be either fiercely rubbed or head-patted. In point of fact, their tactile umwelt is almost certainly different than ours.

First, sensation is not uniform across one's body. Our tactile resolution is different at different points on our skin. We can detect two fingers one centimeter apart at the nape of our necks, but if the fingers are moved down the back we feel that they are touching the same spot. The resolution of touch to animals is likely different still: what we think is a gentle pat may be barely detectable or may be painful.

Second, the somatic—body—map of the dog is not the same as our somatic map: the most sensitive or meaningful parts of the body are different on dogs. As seen in many of the aforementioned agonistic contact actions, grabbing a dog's head or muzzle—the first part a guileless dog-petter reaches for—may be viewed as aggressive. It is similar to what a mother will do to an unruly pup, or an older dominant wolf will do to a member of his pack. Here too are the whiskers (vibrissae), which like all hairs have pressure-sensitive receptors at their ends. The whiskered receptors are specially important to detecting motion around the face or nearby air currents. If you are close enough to see the dog's muzzle whiskers, you might notice them flare when the dog feels aggressive (it might be inadvisable to be so close in that case). Pulling a tail is a provocation, but usually one for play, not aggression—unless you don't let go. Touching the underbelly might prompt a dog to feel sexually frisky, as genital licking often precedes an attempt to mount. A dog rolling over on his back is doing much more than simply revealing his belly: this is the same posture dogs use to allow their mothers to clean their genitals. The forceful belly-rubber may find himself urinated upon.

Finally, just as we have highly sensitive areas—the tip of the tongue, our fingers—so too does the dog. There is a species level to this—no person likes being poked in the eye—and an individual level—I might be ticklish on the bottoms of my feet, while you aren't at all. You can easily do a tactile survey and map your own dog's body. Not only are the favored and prohibited places to touch different, but the very form of contact is crucial. In a dog's world repeated touching is different than constant pressure. Since touch is used to communicate a message, holding a hand in one place on a dog's body conveys that same message writ large. At the same time, full-body contact is preferred by some dogs, especially young dogs, and especially when they are the initiators of the contact. Dogs often find places to lie down that maximize contiguity of body with body. This might be a safe posture for dogs, especially as puppies, when they are entirely reliant on others for their care. To feel light pressure along the whole body is to have assurance of your well-being.

It is hard to imagine knowing a dog but not touching him— or being touched by him. To be nudged by a dog's nose is a pleasure unmatched.

AT HELLO

Early in my life with Pumpernickel, I got a full-time job and she got a classic case of separation anxiety. Mornings as I prepared to leave the house after our walk, she began to whimper, shadow me from room to room, and, finally, vomit. I consulted with trainers who gave me very reasonable guidelines to reduce her stress at separation. I followed all known commonsensical procedures, and before too long Pump returned to a healthy mental and physical state. But there was one dictum I didn't follow. Don't ritualize your departure and later return, they advised; don't celebrate your reunion. I refused. Her snuffly, nosed greeting, our heaping together on the floor in a joyous commemoration of togetherness, was too good to let go.

Lorenz called the greeting between animals after being apart a "redirected appeasement ceremony." That nervous excitement one might feel on suddenly seeing someone else in one's den or territory could lead to two different results: an attack of the potential stranger, or a redirection of the excitement into a greeting. His idea was that there is very little difference between the attack and the greeting, besides a few subtle alterations or additions. Between mallards, one of the birds he studied extensively, two individuals meeting each other engage in a rhythmical "ceremonial to-and-fro movement" that could become aggressive, but for the male mallard, the drake, lifting his head and turning it away. This leads to a mutual ceremony of pretending to preen each other, and the greeting is complete: another fight inhibited.

The greeting among humans is similarly ritualized. We look each other in the eyes, wave hands at each other, hug or kiss once or twice or thrice depending on one's native country. These all may be redirections of an uncertain feeling upon seeing someone else. What is more, we may smile or chuckle. Nothing is more reassuring of the good intent of another person than laughter, Lorenz proposed. This paroxysm of noise is surely most often the expression of joy, but it might also be an eruption typical of alarm reformulated as delight or surprise (not unlike the rough play context in which dog laughter appears).

Having channeled one's excitement into a greeting in this Lorenzian way, one might add other components to the hello. Wolves and dogs do. Their greetings, and the greetings of all social canids, are similar. In the wild, when parents return to the den, the pups mob them, madly lunging at their mouths in the hope of getting them to regurgitate a bit of the kill they have consumed. They lick at their lips, muzzle, and mouth, take a submissive posture, and wag furiously.

As we have seen, what many owners cheerfully describe as "kisses" is face licking, your dog's attempt to prompt you to regurgitate. Your dog will never be unhappy if his kisses in fact prompt you to spit up your lunch. This greeting isn't complete without an excited approach and constant, energetic contact. Ears that were pricked to hear your arrival fall flat against the dog's head, which dips slightly in a submissive gesture. The dog pulls his lips back and drops his eyelids: in humans, markers of a true smile. He wags madly or beats a frantic rhythm with the tip of his tail against the ground. Both wags contain all the excited running-around energy that the dog suppresses in order to stay close to you. He may whine or yelp with pleasure. Adult wolves howl daily: among packs, a chorus of howling may help coordinate their travels and strengthens their attachment. Similarly, if you greet the dog with cries and vocal hellos, your dog may cry back at you. In every move he is breathing and exuding his recognition of you.

If greeting and contact were all, we might expect a rash of monkeys bonded with wolves, of rabbits cohabitating with prairie dogs. They all require contact in infancy. And even ants greet homecomers to the nest. I suppose that, predatory issues aside (a big aside), the potential is there. A gorilla named Koko, taught to use sign language to communicate and raised in a human home, had her own pet kitten. We are relieved of acting instinctually in the way few animals are. But there is one other aspect that makes human-dog bonding unique: timing. We act well together.

THE DANCE

On a long walk Pump stays near me, but not too. If I call her to me, she comes charging forth full-steam and stops just past me. She likes to be one step off. And yet when we walk together on a lean path and she is ahead of me, she *checks*—regularly looking back to see where I am. She only needs to turn her head partway round to see me, lifting it from its regular downward cast, surveying the ground. If I ever lag, she turns all the way round, ears up and attentive: waiting for me. Oh, I love to come to this beckoning stance of hers: I might gallop a bit as I near her, and this cues her to play-bow, or to pivot on her rear legs and assume her trot leading us on our walk.

He has begun, on this second day, to come to a snap: just picked it up right away. We snap him back and forth between us.

Dogs, though they do not hunt cooperatively, are cooperative. Watch the parade of leashed dog-person twosomes along a city street. Despite small diversions, they are dancing in masterful synchrony, traveling *together*. Working dogs are trained to heighten their sensitivity to the dance. Blind people and their guide dogs take turns initiating movement, completing each other.

It helps that dogs live at our speed. A house mouse, its heart beating four hundred times a minute at rest, is always in a hurry; a tick can wait for a month, a year, or eighteen years in suspended animation for that odor of butyric acid to come along; dogs function much more at our pace. Though we outlive them, their lives stretch across a generation. And they *act* at a pace sufficiently close to ours—if slightly quicker—to enable us to discern their movements, imagine their intent. They act in response to our actions, with alacrity. They dance with us.

A puppy initially balks at a leash, pulls at it unyieldingly, or simply fails to grasp that he is tethered to it—and thus to you—as he pulls toward that very interesting newspaper wafting down the sidewalk. In very little time, though, puppies learn to be highly cooperative walking partners, walking at roughly the same rate and often in step with their owners. They *match* their owners, almost mimicking us. In turn, we unconsciously mimic our mimickers. In ethology, this is called "allelomimetic behavior" and is implicated in the development and maintenance of good social relationships among animals. More than that, though, the puppy has learned about the sequence of behaviors that you repeat, that make up a walk—and anticipates them. Before long, he knows the series of steps to get the walk started, the corners you turn on your route to the park, the place where the leash is snapped off or the ball is brought out. He anticipates the long-walk turnaround point; the short-walk turnaround point; and knows how to evade the latter. Some dogs even seem to know exactly how far the parameters of a leash extend from our hands, and they dart about within those parameters, grabbing a stick or sniffing a passing dog without our breaking stride.

Once we take them off their leashes, the dance continues. My conception of the perfect walk, occasionally achieved, has my dog off-leash running not alongside me but in great circles around me, with our average forward progress over the miles more or less the same. Ideally, we encounter a dozen other dogs. There is little as therapeutic as watching two dogs at play together in a boisterous full-bodied brawl: it extends our pleasure at turn-taking games to high-speed, exuberant result. The rules of play—signaling, timing—are similar to our conversational rules. And so we can enter into a dialogue of play with our dogs.

I start it. I inch to where she's lying and I put my hand on her paw. She pulls it away—and puts her paw on my hand. I place my hand over her paw again; more quickly now, she mimics me. We trade slaps like this until it is too much: I laugh, breaking the spell, and she stretches forward over her paws, mouth open nearly in smile, to lick my face. There's a special intimacy of having her put her hand—its weight, the scratchiness of her pads, the feeling of each claw—on mine. Mostly it's the simple fact of the use of this appendage to communicate with me—it is not seen as a hand independent of its arm until she treats it as one, parallel to mine.

The elements that make play enjoyable are hard to pinpoint, just as a great joke always seems to be funnier than its deconstruction. Try getting a robot to play with you: they always seem to lack a certain . . . *playfulness.* A few years ago Sony developed a mechanical pet, "Aibo," designed to look like a dog—it is four-legged, has a tail, characteristic head form, et cetera—and to act something like a dog—it wags, barks, and performs simple trained-dog routines. What the Aibo does not do is play like a dog, and the designers wanted it to be more playfully interactive with people. With this in mind I studied dogs and humans playing together: wrestling, chasing, tossing and retrieving balls and sticks and ropes. I watched, videotaped, and then transcribed all the behaviors that each of the participants did. Then I looked for the elements that were consistent across the successful bouts of this interspecies play.

What I hoped to find were clear routines and games that could be modeled in a doggish toy such as Aibo. What I found was both simpler and more powerful. In every bout, the player's actions were importantly *contingent* on—based on and related to—the other's actions. This established a rhythm to the play.

Such contingency is easily seen in even very early human social interaction. At two months, infants coordinate simple movements with their mothers, such as mirroring facial expressions. In play, coordinated responses to actions, such as a ball leaving a thrower's hand, happened in as little as five frames of the videotape (approximately one-sixth of a second). Mirrored responses—lunging after being lunged at, for example—are rife during play. The timing is crucial: dogs respond to our movements in the time frame another human might.

A simple game of fetch, for instance, is a dance of call and response. We enjoy the game because of the dog's reactive readiness to respond to our actions. Cats, by contrast, are simply not enjoyable fetch playmates: they may in fact fetch you an object, but in their own time. Dogs participate in a kind of communion with their owners around the ball, with each responding at a conversational pace: in seconds, not hours. The dogs are acting like very cooperative humans. Another game is simply doing an activity in parallel: running together. In play between dogs parallelism is common. Two dogs may mimic each other's gaping mouths yawing back and forth. Often one dog will observe and then match the other's preoccupation: hole digging, stick chewing, ball trumpeting. As wolves hunt together collaboratively, this ability to act with others, matching their behavior, might come from their ancestry. To have your play-slap matched by a dog's is to feel suddenly in communication with another species.

We experience the dog's responsiveness as expressive of a mutual understanding: we're on this walk *together;* we're playing *together.* Researchers who have looked at the temporal pattern of interactions with our dogs find that it is similar to the timing patterns among mixed-sex strangers flirting, and to the timing among soccer players as they move down the field that feels like great teamwork. There are hidden sequences of paired behaviors that repeat in interaction: a dog looking at the owner's face before picking up a stick, a person pointing and a dog following the point to what it's directed. The sequences are repeated, and they are reliable, so we begin to get the feeling, over time, that there is a shared covenant of interaction between us. None of the sequences is itself profound, but none is random, and together they have a cumulative result.

Walk down Fifth Avenue in midtown Manhattan around lunchtime on a weekday and you experience the frustration and pleasure of being a member of the human species. The sidewalks are mobbed, jammed with tourists wandering and gawking; office workers rushing to grab lunch or dallying before returning; enterprising street vendors rushing from enforcement officers. It is a formidable sight, one you may not relish joining. On most days, though, you can take any pace you'd like, and just as easily wend your way through the crowds. It has been speculated that people walking en masse don't crash into each other because we are instantly and easily predictable. It only takes a glance to calculate when the oncoming person will reach you. You unconsciously veer subtly right to avoid him; he has done the same with you. It is not unlike (but not quite as completely successful as) the school of fish that abruptly, with one mind, turns tail and goes back from where it came. We are social, and social animals coordinate their actions. What dogs do is cross the species line and coordinate with us. Pick up the leash of any dog in your neighborhood and suddenly you are walking together, like old friends.

The significance of these three elements is corroborated by the kinds of feelings generated when they disappear: of mild betrayal, of momentary severance of the bond. There's a feeling of disconnect when a dog one reaches for ducks her head away, preventing contact. The frustration is immediate when a dog stops cooperating in taking turns in a game: refusing to bring the ball back, not seeing the toss or pursuing a seen toss. A betrayal is felt when the simple communication *come!* isn't followed by a dog *coming*. And it would be heartbreaking to approach your dog and to fail to prompt a tail to wag, ears to flatten to the head, or a stomach to be bared for scratching. Dogs whom we perceive as stubborn or disobedient are those dogs who flout these elements. But these elements are natural for both them and for us; a disobedient dog more likely simply does not realize what rules he is being asked to obey.

THE BOND EFFECT

Our bond with dogs is strengthened by contact, by synchrony, and by marking reunions with a greeting ceremony. So too are we strengthened by the bond. Simply petting a dog can reduce an overactive sympathetic nervous system within minutes: a racing heart, high blood pressure, the sweats. Levels of endorphins (hormones that make us feel good) and oxytocin and prolactin (those hormones involved in social attachment) go up when we're with dogs. Cortisol (stress hormone) levels go down. There is good reason to believe that living with a dog provides the social support which correlates with reduced risk for various diseases, from cardiovascular disease to diabetes to pneumonia, and better rates of recovery from those diseases we do get. In many cases, the dog receives nearly the same effect. Human company can lower a dog's cortisol level; petting can calm a racing heart. For both of us, this is a kind of placebo, which is not to say that it isn't real, but that a change is induced in us without a known agent of the change. Bonding with a pet can do the work that long-term use of prescribed drugs or cognitive behavioral therapy do. Of course, it can go wrong, too: separation anxiety is the consequence of a dog feeling so very attached that he cannot stand a moment of detachment.

What are the other results of the bond? We've seen how much they know about us—our smell, our health, our emotions—due not just to their sensory acuity but also to their simple familiarity with us. They come to know how we normally act, smell, and look over the course of our days, and then they are able to notice, many times in ways we cannot, when there is a deviation. The bond effect works because dogs are, at their best, acting as extremely good social interactants. They are responsive, and, crucially, they pay attention to us.

And this connection to us runs deep. A simple experiment consisting of dogs and yawning humans indicates that our link is instinctual—on the level of reflex. Dogs catch our yawns. Just as happens between humans, dog subjects who saw someone yawning themselves began uncontrollably yawning in the next few minutes. Chimpanzees are the only other species we know of for whom yawning is contagious. Spend a few minutes yawning at your own dog (trying not to glare, giggle, or give in to his inevitable complaints) and you can see for yourself this deep-seated connection between human and dog.

Yawning dogs aside, there is a limit to the science here. Science is quite intentionally not looking at the very feature that is most important to dog owners: the feel of the relationship between person and dog. That feel is made up of daily affirmations and gestures, coordinated activities, shared silence. It can be deconstructed somewhat with the dull butter knife of science, but it cannot be reproduced in an experimental setting: it is importantly non-experimental. Experimenters often use what is called a *double-blind* procedure to assure the validity of their data. The subject is always blind to the point of the experiment, and in a double-blind the experimenter is also blind to which subject's data—one from the experimental group or the control group—he is analyzing. In that way, one avoids inadvertently seeing a subject's behavior as fitting in just a little more tightly with the tested hypothesis.

Dog-human interactions, by contrast, are happily double-seeing. We have the feeling of knowing exactly what the dog is doing; the dog may, too. What we think we see is not the stuff of good science, but it is the stuff of a rewarding interaction.

The bond changes us. Most fundamentally, it nearly instantly makes us someone who can commune with animals—with this animal, this dog. A large component of our attachment to dogs is our enjoyment of being *seen* by them. They have impressions of us; they see us in their eyes, they smell us. They know about us, and are poignantly and indelibly attached to us. The philosopher Jacques Derrida ruminated on his cat seeing him nude: he was startled and embarrassed. To Derrida, what was startling was that the animal reflected his image back to him. When Derrida saw his cat, what he saw was his cat *seeing him,* in nakedness.

He was right to implicate our self-regard in our regard of our pets. (As far as I know, though, Derrida never had a dog: his discomfiture might have been greater at the dog's superior gaze.) Of course we revel in the animals themselves. Still, part of what we see when we look at a dog is: the dog looking at us. This is a component of our bond, too. I still imagine my own dog, Pumpernickel, looking at me, seeing herself in my eyes. And I look at her, seeing myself in hers.

The Importance
of Mornings

Pump changed my own umwelt. Walking through the world with her, watching her reactions, I began to imagine her experience. My enjoyment of a narrow winding path in a shady forest, lined with low bushes and grasses, comes in part from seeing how Pump enjoyed it: the cool of the shade, of course, but also the *pathiness,* allowing her to zoom along unchecked, stopping only for rousing scents along the sides.

I now see city blocks, and their sidewalks and buildings, with their investigatory sniffing possibilities in mind: a sidewalk along an uninterrupted wall without fences, trees, or variation, is a block I'd never want to walk down. Where I'll choose to sit in the park—which bench, what rock—is based on where a dog at my side would have the best panoramic olfactory view. Pump loved large open lawns—to plop down in, to roll repeatedly in, to sniff endlessly—and high grass or brush—to lope regally through. *I* came to love large open lawns and high grass and brush in anticipation of her enjoyment. (The interest in rolling in unseen smells remains elusive . . .)

I smell the world more. I love to sit outside on a breezy day. My day is tilted toward morning. The importance of mornings has always been that if I awoke early enough, we could have a long, off-leash walk together in a relatively unpeopled park or beach. I still have trouble sleeping in.

It is a very small bit comforting to realize how deeply she is in me, even over a year from the day when she was also aside me, willing to submit to a tickle of the dense curls under her chin as she rested it on the ground for the last time.

Sitting with a dog on my lap, considering what we know about dogs' abilities, experiences, and perception, I feel partway to full dogness myself. Also, right now, I am covered with dog hair.

Even without getting coated with fur, the knowledge of dog science brings us closer to an understanding of, and appreciation for, dog behavior: how it arises from the ancestral canid, from domestication, from their sensory acuteness, and from their sensitivity to us. With any luck it will get under your skin and you will see the dog from the dog's point of view. Along the way, here is a smattering of ideas of umwelt-ful ways of relating to your dog, of interpreting their behavior, and of considering them in our lives.

GO FOR A "SMELL WALK"

Most of us would agree that we go for walks with dogs for the dog's sake. It is for Pump's sake that I woke early every morning, to catch a permitted off-leash walk in the park; for her sake that I came home during the day to circle the block with her; for her sake that I shod myself before bed and sleepwalked a walk. Yet dog-walks are often not done with the dog's sake in mind, but strangely playing out a very human definition of a walk. We want to make good time; to keep a brisk pace; to get to the post office and back. People yank their dogs along, tugging at leashes to get noses out of smells, pulling past tempting dogs, to get on with the walk.

The dog doesn't care about making good time. Instead, consider the walk your dog wants. Pump and I had a good variety. There were the smell walks, where we made zero progress but she inhaled untold purple, mesmerizing molecules. There were Pump's-choice walks, where I let her choose which way we went at every intersection. There were serpentine walks, where I restrained myself instead of her as she weaved on leash from my left to my right and back again. As a younger dog, she tacitly agreed to go on runs with me when I agreed to occasionally stop and circle around her as she circled around an interesting dog. As she got older, there were even non-walking walks, where she lay down, and just stayed put until she was ready to move on.

TRAIN THOUGHTFULLY

Teach your dog the things you want in a way he can understand: be clear (about what you want him to do), consistent (in what you ask and how you ask it), and tell him when he has got it right (reward him straightaway and often). Good training comes from understanding the mind of a dog—what he perceives and what motivates him.

Avoid the missteps common to those who have the classic idea of what a dog should do: sit, stay, obey. Your dog is not born knowing what you mean by *come here*. You must teach it explicitly, in small steps, and reward him when he actually comes. Dogs are attuned to tiny cues from you, cues that may be the same when you call *come* as when you say *go away!*: a tone of voice, a body posture. It is up to you to make your request specific and distinctive.

Training can take a long time; be patient. When even a "trained" dog does not come to calling, too often people chase him down and then punish him—forgetting that from the dog's point of view, the punishment is linked with your arrival, not his earlier disobedience. This is a quick, effective way to get him to never come when you call him.

When *come here* has been learned, a good argument can be made that there is little else by way of commands that an ordinary dog needs to know. Teach them more if you both enjoy it. What a dog most needs to learn is the importance of you—and that is something he is born to see. A dog who cannot "shake hands" on command is just a little more doggy. Make clear what behaviors you dislike and be consistent in not reinforcing them. Few celebrate a dog who jumps at people as they approach—but start with the premise that it is we who keep ourselves (and our faces) unbearably far away, and we can come to a mutual understanding.

ALLOW FOR HIS DOGNESS

Let him roll in *whatever-that-thing-is* once in a while. Endure some traipsing through mud puddles. Walk off-leash when you can. When you cannot walk off-leash, do not yank him along by his neck, ever. Learn to distinguish a nip from a bite. Let approaching dogs smell each other's rumps.

CONSIDER THE SOURCE

Why does he do that? I am asked almost daily. Many times my only answer can be that not every behavior a dog does has an explanation. Sometimes when a dog suddenly flops on the ground and looks at you, he is just *lying down and looking*—and nothing more. Not every behavior signifies something. Those that do mean something should be explained by taking into consideration the natural history of your dog—as an animal, as a canid, and as a particular breed.

Breed matters: A dog that stares down invisible prey or slowly stalks other dogs may be presenting very good "eye" behavior for a herder. So too with the dog who is aggrieved when one person leaves the room or who nips at everyone's heels as they wander down the hallway. Freezing at movement in the bushes slows down your walk, but it is very good pointing behavior. A bred dog with no task may be agitated, restive, keyed up: a drifter, not clearly driven to any activity. Give him some. This is the great science behind "tossing a ball": a retriever is made happy just to do it, over and over. He is fulfilling his capability. On the other hand, if your dog has a short nose and has trouble breathing, don't assume he can run with you. That same dog, with his near, central vision, may not care for the game *fetch,* while a retriever with a wide visual streak may care only for it. Give your dog a context to play out his innate tendencies—and indulge him a little staring at the bushes now and then.

Animalness matters: adapt to your dog's capacities rather than simply expecting him to adapt to our strange notions of how to be a dog. We want our dogs to *heel*—I have seen people turn furious when their dog does not—but dogs may be more or less prone to walk close to, and in step with, their social companions. Retrievers do, but sporting breeds might not (both will keep an eye on you). Also, most dogs exhibit handedness—*pawedness*—so while we shunt them to our left, as every training class has us do, we might be disadvantaging some dogs more than others (and leading to inevitable frustration if the good smells are all on the right side of the path). It would be a shame to punish a dog needlessly because we simply do not know his nature. Not every dog needs to heel in the same way: the essence is simply being safe and manageable.

Canidness matters: Your dog is a social creature. Do not leave him alone for most of his life.

GIVE HIM SOMETHING TO DO

One of the best ways to see your dog's capacities and interests is simply to provide a lot of possible things to interact with. Wiggle a string in front of your dog's nose along the ground; stash a treat in a shoebox; or invest in the many creative dog toys that are marketed. A rich set of things to burrow into, nose, chew, bob, shake, pursue, or watch will engage your dog—and keep him from finding his own burrowable and chewable objects among your possessions. Outside, agility training or some simulacrum obstacle course is a well-defined way to engage and interest many energetic but driven dogs. But interest can be spiked simply by a weaving, smell-laden path, or the unexplored reaches of a field.

Dogs like both the familiar and the new. Happiness is novelty—new toys, new treats—in a safe, well-known place. It can be cure for boredom, too: the new requires attention and prompts activity. Hiding food to be searched for is one example: they must move around to explore the space, using nose and paw and mouth together. You need only watch the exuberance of an agility dog on a new course to see how good *new* is.

PLAY WITH HIM

In youth, but even throughout their lives, dogs are constantly learning about the world, like the developing child. Games that children find mind-bogglingly fun work with dogs too. Peeka-boo, disappearing around the corner or under a blanket instead of behind hands, is especially fun when dogs are learning about invisible displacement, that objects continue to exist when you can no longer see them. Dogs are astute perceivers of associa-tions, and you can play with that: ring the bell before dinner, Ivan Pavlov found, and the dogs anticipate dinner. You can connect bells—or horns, whistles, harmonica, gospel music, just about anything—not just with food, but with people arriving, or the time for a bath. Make a string of associations—and treat your dog's actions as adding to that string. Play imitation games, mirroring what your dog does: jumping on the bed, yelping, pawing the air. Note your dog's current skills, and try to stretch his ability. If he seems to know *walk* or *ball,* start using words that make more subtle distinctions: *smell walk* and *blue ball; evening smell walk* and *blue squeaky ball.* And at any age, play with your dog as a dog would. Choose your play signal—play-slap your hands on the ground, mimic panting close to his face, race away with looks back at him—and play. Treat your hands as he does his mouth and grab head, legs, tail, belly. Give him a good toy to hold on to, or be prepared for some nips. Watch as your own tail may begin to wag.

LOOK AGAIN

Much enjoyment can be had in noticing the invisible-visible features of your dog: the things we typically see through that are on display right in front of us. We now know how attentive dogs can be to people, and to our attention: notice the various and creative methods your dog uses to try to get your attention. Does he bark or bray? Stare at you wistfully? Sigh loudly? Walk back and forth between you and the door? Lay his head on your lap? Find the methods you like, and respond to them, letting the others fade away naturally.

Notice how your dog uses his eyes; the frenzy of his nose; how his ears fold back, prick up, and pivot toward a distant bark. Notice all the sounds he makes, and all the sounds he notices. Even the way the dog moves, an action so familiar as to make him recognizable at a distance, is transformed on closer examination: what gait does he use? A medium-sized dog may stride forward in the classic *walk,* the rear foot on one side of the body slowly chasing the front foot to the ground, the diagonal paws moving almost in sync. Hurrying a little, he *trots,* the diagonal legs now in tandem, occasionally finding himself with only one paw of four fully on the ground. Between the trot and the walk is the gait of the short-legged: typical of the bulldog, front-heavy with a wide stance, his rear end rolling as he walks. Leggy dogs do better at the *gallop,* the run of greyhounds, wherein the two rear feet precede the two front to the ground, the dog's body alternating between outstretched, and airborne and spring-loaded. In the gallop that fifth toelike digit partway up the front leg of most dogs—the dew claw—is used for stability and leverage; at a gallop's end you might find the usually clean dew claw with a dollop of mud under it. Toy-sized dogs *half-*

hound, bringing their two hind legs forward at once but uncoupling their front footfalls. Other dogs *pace,* their left legs moving forward and falling at once, followed quickly by the right. Mesmerize yourself trying to keep track of the complexity of your dog's gait.

SPY ON HIM

To understand what your dog's day at home without you might be like, by all means videotape it. One of the distinct pleasures I got with Pumpernickel was seeing her act without me. Despite hours of videorecording, I rarely turned my camera on to her. It was only when she didn't expect me—when a friend had taken her out, and I arrived unannounced—that I got to see her carry on without me.

It was spectacular to see. You can re-create this kind of spectacle by setting up a videotape at your home when you leave for the day. I recommend this "eavesdropping" not because it reliably reveals spectacles—it does not—but because it allows you to see your dog's life without you there. You will more fully understand what his day might be like by watching a snippet of the day pass later, minute by minute.

What I saw in my eavesdropping was Pump's independence, freed not just from the need to check back with me, but from the kind of scrutiny to which I subjected all her behavior. She existed capably without me, for the hours that I milled about in the bookstore, had an extra-long run, went elsewhere for dinner followed by elsewhere still for drinks. This was at once reassuring and fully humbling. I am glad that she managed the day on her own, yet I am sometimes mystified that I ever left her alone at all.

Most dogs are simply alone all day with little to do, expected to wait it out until we return, and then act just as we want them to. And we are surprised and horrified when they actually do something in our absence! That dogs endure this (and much worse misinterpretation and neglect) is almost part of their constitution. We can, and do, get away with it. But dogs are individuals. It is for this reason that they require—and deserve—more attention to their umwelt, to their experience, to their point of view.

DON'T BATHE YOUR DOG EVERY DAY

Let them smell like a dog as long as you can stand it. Some dogs will even develop painful skin sores from regular bathing. And no dog wants to smell like a bathtub that has had a dog in it.

READ THE DOG'S TELLS

Like novice poker players, dogs reveal what could be called their "tells"—their intent, their "hand"—with every move, if you simply look. The configuration of the face, head, body, and tail are all meaningful. And there is more to it than whether the tail is wagging or the dog is barking: dogs can say more than one thing at a time. A barking dog whose tail fans the sky is not "about to attack" but is instead more curious, alert, uncertain— and interested. A furiously wagging low tail undermines the aggressiveness of a familiar dog snarling as he guards a ball.

Given the salience of eye contact to all canids, and the dog's use of gaze, you can get a lot of information about an unknown dog from his eyes. Constant eye contact can be threatening: do not approach a dog by gazing non-stop, which may be perceived as staring him down. If he is staring at you, you can deflect his gaze by turning away slightly, breaking eye contact. They do the same when they are tense: turning their head to the side, or distracting themselves with a yawn or a sudden interest in a smell on the ground. If you think you are the recipient of a threatening stare, you can confirm it by looking for its accompaniment: hackles up, ears up, tail up, body frozen. A stare with a tongue darting licks into the air is more adoring than aggressive.

PET FRIENDLY

Though they nearly all look pettable, not every dog likes to be petted. Attending to that is not only polite, it is sometimes imperative: a fearful or sick dog might respond to touch with aggression. There are great individual differences in dogs' sensitivity to petting, and their current interest can be changed by their state of health, their state of happiness, and past experience. For most dogs, the right touch by a human is a calming, bonding experience. Light touch is irritating or exciting; a firm hand is relaxing; too firm a hand is probably oppressive. They (and you) can be physically calmed by steady, continuous strokes from the head to rump, or by capable deep muscle massage. Watch your dog's reaction and find his preferred touch zones. And let him touch you back.

GET A MUTT

If you don't have a dog yet, or are getting another dog, I have just the breed for you: the breedless dog, the mutt. The myth that a shelter dog, especially a mixed-breed dog, will be less good or less reliable than a purebred dog is not just wrong, it is entirely backward: mixed breeds are healthier, less anxious, and live longer than purebreds. When you buy a bred dog, you are simply not buying a fixed object, guaranteed to act in certain ways—regardless of what the breeder tells you. What you might get is a dog with an overriding fixation, born of breeding for a task that he will likely never do while living with you (who nonetheless will still be wonderfully doggy). Mutts, on the other hand, with the bred characteristics diluted, wind up having lots of latent abilities and less mania.

ANTHROPOMORPHIZE WITH UMWELT IN MIND

On walks Pump was never satisfied with being on one side of the path or the other: she weaved back and forth capriciously. Holding her on leash I was constantly readjusting my hand on the thing. Sometimes I'd insist she stay to one side of me, and she sighed at me while we both glanced knowingly at the good un-smelled spots on the other side.

Even with a scientific take on the dog, we find ourselves using anthropomorphic words. Our dogs—my dog—make friends, feel guilty, have fun, get jealous; understand what we mean, think about things, know better; are sad, are happy, are scared; want, love, hope.

This way of talking is easy, and sometimes useful, but it is also part of a bigger, more exceptionable phenomenon. As we recast every moment of a dog's life in human terms, we have begun to completely lose touch with the animal in them. It is no longer the rare dog who is shampooed, clothed in garments, and feted on his birthday. That may seem benign, but it is also part of a de-animalizing of dogs that is somewhat radical. We are rarely present for their births, and many people will choose not to be present for their own dogs' deaths. We eliminate sex for the most part: we neuter dogs and we discourage the slightest lascivious thrusts of the hips. They are fed sanitized food, in bowls; they are largely restricted to a leash-length distance from our heels. In cities, their excrement is bundled up and thrown away. (Happily, we have not yet taught them to use the toilet . . . convenient though we know that would be.) Breed types are described like products, with specified features. It seems as though we are trying to get rid of the animal part of the dog.

If we assume that we have reduced the animal factor to zero, we are in for some unhappy surprises. Dogs do not always behave just as we think they should. They may sit, lie down, and roll over—but then will revert magnificently. They suddenly squat and urinate in the house, bite your hand, sniff your crotch, jump on a stranger, eat something gnarly in the grass, don't come when you call them, roughly tackle a much smaller dog. In this way, our frustrations with dogs often arise from our extreme anthropomorphizing, which neglects the very animalness of dogs. A complex animal cannot be explained simply.

The alternative to anthropomorphizing is not simply treating animals as precisely unhuman. We now have the tools to take a more measured look at their behavior: with their umwelt and their perceptual and cognitive abilities in mind. Nor need we take a dispassionate stance toward animals. Scientists anthropomorphize . . . at home. They name their pets, and see love in a named-dog's upward-turned gaze. In research, names are verboten: while they might help tell animals apart, they are not benign. Naming a wild animal "colors one's thinking about it forever afterwards," a preeminent field biologist noted. There are obvious observational biases that are introduced when you name the subject of your observations. Jane Goodall famously violated this maxim, and "Graybeard" became known to the world. But "Graybeard" for me connotes a wise, old man: as a result, I may be more likely to perceive his behavior as indicative of his wisdom than see it as foolishness. Instead, to distinguish individual animals, most ethologists use identifying markings—leg bands, tags, or marking fur or feathers with dye—or look for identity in habitual behavior, social organization, or natural physical features.*

To name a dog is to begin to make him personal—and thus an anthropomorphizable creature. But we must. To name a dog is to assert an interest in understanding the nature of the dog; to not name the dog seems the pinnacle of disinterest. Dogs named *Dog* make me sad: the dog is already defined out of being a player in the owner's life. *Dog* has no name of his own; he is only a taxonomic subspecies. He will never be treated as an individual. What one is doing when naming a dog is starting him on the personality that he is to grow into. When trying out names for our dog, calling words out at her—*"Bean!" "Bella!" "Blue!"*—to see if any prompted a reaction, I felt that I was

---

*Neither are these methods benign, in some cases: there is the famous case of the zebra finches, captured and harmlessly leg-banded for identification as the researchers observed their mating tactics. Lo and behold, the only feature that they found was predictive of a male's success at breeding was the color of his leg band. Female zebra finches apparently swoon for a red band on a fella (males prefer black-banded females).

searching for "her name": the name that was already hers. With it, the bond between human and animal—wrought of understanding, not projection—could begin to form.

Go look at your dog. Go to him! Imagine his umwelt—and let him change your own.

Postscript:
Me and My Dog

I sometimes find deep recognition in photos of her in which her eyes were not distinguishable from the darkness of her coat. It represents to me the way in which there was always something mysterious about her existence to me: what it was like to be Pump. She never laid it out there in the open. She had a privacy about her. I feel privileged that I was let into that private realm.

Pumpernickel wagged into my life in August 1990. We spent nearly every day together, until the day of her last breaths in November 2006. I still spend every one of my days with her.

Pump was a total surprise. I didn't expect to be changed constitutionally by a dog. But it quickly became apparent that the description "a dog" didn't capture the astounding abundance of facets to her, the depth of her experience, and the possibilities of a lifetime knowing her. Before long, I felt pleasure simply at her company and pride at watching her act. She was spirited, patient, willful, and disarming all in one great furry bundle. She was sure of her opinions (she had no truck with yipping dogs) and yet open to new things (as the occasional fostered cat—despite each's unwavering disinterest). She was effusive; she was responsive; she was great fun.

What Pump was not was a subject in my research (at least, not intentionally). Still, I brought her with me when going to watch dogs. She was often my passkey into dog parks, and into dog circles: without a dog companion, a person may be treated suspiciously by dogs and owners alike. As a result, she wanders through many of my videos of bouts of play—through and out, as my video camera was trained on my unwitting subjects, not on my Pump. Now I regret my camera's unemotional oversight of her. Though I captured the social interactions I wanted, and was eventually, after much reviewing and analysis of the behaviors of the interactants, able to discover some surprising abilities of dogs, I missed some moments of *my* dog.

Every dog owner would agree with me, I suspect, about the specialness of her own dog. Reason argues that everyone must be wrong: by definition, not every dog can be the special dog—else special becomes ordinary. But it is reason that is wrong: what is special is the life story that each dog owner creates with and knows about his own dog. I am not exempt from feeling that, even from a scientific vantage. Behavioral scientific approaches to dogs, far from displacing this story, simply build on the singular understanding of the dog owner—on the expertise that each dog owner has about her dog.

When Pump was nearly at the end of her life, undeniably old, she lost weight, her muzzle grayed, she slowed sometimes to a stop on walks. I saw her frustrations, her resignations, her impulses pursued or abandoned; I saw her considerations, her control, her calm. But when I looked at her face, and into her eyes, she was a puppy again. I saw glimpses of that unnamed dog who so cooperatively let us plop a too-big collar around her neck and walk her out of the shelter and thirty blocks home. And then thousands of miles since.

After knowing Pump, and losing Pump, I met Finnegan. I already cannot imagine not knowing this new character: this leaner on legs, this stealer of balls, this warmer of laps. He is incredibly unlike Pumpernickel. Yet what she taught me has made every moment with Finnegan infinitely richer.

She lifted her head and turned toward me, her head pulsing slightly with her breathing. Her nose was dark and wet, her eyes calm. She began licking, full long licks of her front legs, of the floor. The tags of her collar clonked on the wood. Her ears lay flat, curling up a little at the bottom like a felted leaf, dried in the sun. Those days her front toes were a little splayed, her paws turned clawlike as though in preparation to pounce. She did not pounce. She yawned. It was a long, lazy afternoon yawn, her tongue languidly examining the air. She settled her head down between her legs, exhaled a kind of *har-ummmp,* and closed her eyes.

Notes and Sources

In addition to the sources listed by chapter below, I refer to the following books frequently. Each is a scholarly yet accessible approach to dog behavior, cognition, or training; I recommend them all for anyone interested in further details of dog science.

Lindsay, S. R. 2000, 2001, 2005. *Handbook of applied dog behavior and training* (3 volumes). Ames, Iowa: Blackwell Publishing.

McGreevy, P., and R. A. Boakes. 2007. *Carrots and sticks: Principles of animal training*. Cambridge: Cambridge University Press.

Miklósi, Á. 2007. *Dog behavior, evolution, and cognition*. Oxford: Oxford University Press.

Serpell, J., ed. 1995. *The domestic dog: Its evolution, behaviour and interactions with people*. Cambridge: Cambridge University Press.

PRELUDE

on determining species brain differences:

Rogers, L. 2004. Increasing the brain's capacity: Neocortex, new neurons, and hemispheric specialization. In L. J. Rogers, and G. Kaplan, eds. *Comparative vertebrate cognition: Are primates superior to non-primates?* (pp. 289–324). New York: Kluwer Academic/Plenum Publishers.

UMWELT

on the dolphin smile:

Bearzi, M., and C. B. Stanford. 2008. *Beautiful minds: The parallel lives of Great Apes and dolphins.* Cambridge, MA: Harvard University Press.

on the fear grin in chimpanzees:

Chadwick-Jones, J. 2000. *Developing a social psychology of monkeys and apes.* East Sussex, UK: Psychology Press.

on eyebrow-raising in monkeys:

Kyes, R. C., and D. K. Candland. 1987. Baboon (*Papio hamadryas*) visual preferences for regions of the face. *Journal of Comparative Psychology, 4,* 345–348.

de Waal, F. B. M., M. Dindo, C. A. Freeman, and M. J. Hall. 2005. The monkey in the mirror: Hardly a stranger. *Proceedings of the National Academy of Sciences, 102,* 11140–11147.

on chicken preferences:

Febrer, K., T. A. Jones, C. A. Donnelly, and M. S. Dawkins. 2006. Forced to crowd or choosing to cluster? Spatial distribution indicates social attraction in broiler chickens. *Animal Behaviour, 72,* 1291–1300.

on muzzle biting and standing-over in wolves:

Fox, M. W. 1971. *Behaviour of wolves, dogs and related canids.* New York: Harper & Row.

on shock experiments:

Seligman, M. E. P., S. F. Maier, and J. H. Geer. 1965. Alleviation of learned helplessness in the dog. *Journal of Abnormal Psychology, 73,* 256–262.

on umwelt, ticks, and functional tones:

von Uexküll, J. 1957/1934. A stroll through the worlds of animals and men.

In C. H. Schiller, ed. *Instinctive behavior: The development of a modern concept* (pp. 5–80). New York: International Universities Press.

on pessimistic rats:

Harding, E. J., E. S. Paul, and M. Mendl. 2004. Cognitive bias and affective state. *Nature, 427,* 312.

on dog kisses:

Fox, 1971.

on the dog's sense of taste:

Lindemann, B. 1996. Taste reception. *Physiological Reviews, 76,* 719–766. Serpell, 1995.

"dogs have . . . a striking way of exhibiting their affection . . ."

Darwin, C. 1872/1965. *The expression of the emotions in man and animals.* Chicago: University of Chicago Press, p. 118.

BELONGING TO THE HOUSE

on the variety of canids:

Macdonald, D. W., and C. Sillero-Zubiri. 2004. *The biology and conservation of wild canids.* Oxford: Oxford University Press.

on raisin toxicity:

McKnight, K. Feb. 2005. Toxicology brief: Grape and raisin toxicity in dogs. *Veterinary Technician, 26,* 135–136.

etymology of "domesticated":

I drew this wording from Samuel Johnson's 1755 dictionary: *domestical* and *domestick* are both partially defined "belonging to the house; not relating to things publick."

on the fox domestication experiments:

Belyaev, D. K. 1979. Destabilizing selection as a factor in domestication. *Journal of Heredity, 70,* 301–308.
Trut, L. N. 1999. Early canid domestication: The farm-fox experiment. *American Scientist, 87,* 160–169.

on wolf behavior and anatomy:

Mech, D. L., and L. Boitani. 2003. *Wolves: Behavior, ecology, and conservation.* Chicago: University of Chicago Press.

on domestication:

There are many current theories of dog domestication. The one presented here is corroborated by both the recent mtDNA findings, and by a better understanding of the genetics of selection. It is elaborated in R. Coppinger and L. Coppinger. 2001. *Dogs: A startling new understanding of canine origin, behavior, and evolution.* New York: Scribner.
Clutton-Brock, J. 1999. *A natural history of domesticated mammals,* 2nd ed. Cambridge: Cambridge University Press.

on earliest date of domestication:

Ostrander, E. A., U. Giger, and K. Lindblad-Toh, eds. 2006. *The dog and its genome.* Cold Spring Harbor, NY: Cold Spring Harbor Laboratory Press.
Vilà, C., P. Savolainen, J. E. Maldonado, I. R. Amorim, J. E. Rice, R. L. Honeycutt, K. A. Crandall, J. Lundeberg, and R. K. Wayne. 1997. Multiple and ancient origins of the domestic dog. *Science, 276,* 1687–1689.

on development:

Mech and Boitani, 2003.
Scott, J. P., and J. L. Fuller. 1965. *Genetics and the social behaviour of the dog.* Chicago: University of Chicago Press.

on poodle/husky difference in development:

Feddersen-Petersen, D., in Miklósi, 2007.

on wolf rope task:

Miklósi, Á., E. Kubinyi, J. Topál, M. Gácsi, Zs. Virányi, and V. Csányi. 2003. A simple reason for a big difference: Wolves do not look back at humans, but dogs do. *Current Biology, 13,* 763–766.

on eye contact:

Fox, 1971.
Serpell, J. 1996. *In the company of animals: A study of human-animal relationships.* Cambridge: Cambridge University Press.

on breeds:

Garber, M. 1996. *Dog love.* New York: Simon & Schuster.
Ostrander et al., 2006.

on leg length-chest depth ratios:

Brown, C. M. 1986. *Dog locomotion and gait analysis.* Wheat Ridge, CO: Hoflin Publishing Ltd.

on the Ibizan and Pharaoh:

Parker, H. G, L. V. Kim, N. B. Sutter, S. Carlson, T. D. Lorentzen, T. B. Malek, G. S. Johnson, H. B. DeFrance, E. A. Ostrander, and L. Kruglyak. 2004. Genetic structure of the purebred domestic dog. *Science, 304,* 1160–1164.

on breed specs:

Crowley, J., and B. Adelman, eds. 1998. *The complete dog book,* 19th edition. Publication of the American Kennel Club. New York: Howell Book House.

on the dog genome:

Kirkness, E. F., et al. 2003. The dog genome: Survey sequencing and comparative analysis. *Science, 301,* 1898–1903.
Lindblad-Toh, K., et al. 2005. Genome sequence, comparative analysis and haplotype structure of the domestic dog. *Nature, 438,* 803–819.

Ostrander et al., 2006.
Parker et al., 2004.

on aggressive breeds:

Duffy, D. L., Y. Hsu, and J. A. Serpell. 2008. Breed differences in canine aggression. *Applied Animal Behavior Science, 114,* 441–460.

on sheepdog behavior:

Coppinger and Coppinger, 2001.

on packs:

Mech, L. D. 1999. Alpha status, dominance, and division of labor in wolf packs. *Canadian Journal of Zoology, 77,* 1196–1203.
Mech and Boitani, 2003, especially L. D. Mech, and L. Boitani. "Wolf social ecology" (pp. 1–34) and Packard, J. M. "Wolf behavior: Reproductive, social, and intelligent" (pp. 35–65).

on dog and wolf tracks:

Miller, D. 1981. *Track Finder.* Rochester, NY: Nature Study Guild Publishers.

on feral dogs:

Beck, A. M. 2002. *The ecology of stray dogs: A study of free-ranging urban animals.* West Lafayette, IN: NotaBell Books.

on Italian free-ranging dogs:

Cafazzo, S., P. Valsecchi, C. Fantini, and E. Natoli. 2008. Social dynamics of a group of free-ranging domestic dogs living in a suburban environment. Paper presented at Canine Science Forum, Budapest, Hungary.

on the wolf socialization project:

Kubinyi, E., Zs. Virányi, and Á Miklósi. 2007. Comparative social cognition: From wolf and dog to humans. *Comparative Cognition & Behavior Reviews, 2,* 26–46.

"blooming, buzzing confusion":

William James used these words to describe the lack of organization of the information that an infant first receives through her inchoate senses: James, W. 1890. *Principles of psychology.* New York: Henry Holt & Co., p. 488.

"white and shapeless lump of flesh . . .":

Pliny the Elder. *Natural history* (tr. H. Rackham, 1963), Volume 3. Cambridge, MA: Harvard University Press, Book 8(54).

SNIFF

generally interesting readings on smell:

Drobnick, J., ed. 2006. *The smell culture reader.* New York: Berg.
Sacks, O. 1990. "The dog beneath the skin." In *The man who mistook his wife for a hat and other clinical tales* (pp. 156–160). New York: HarperPerennial.

on sniffing:

Settles, G. S., D. A. Kester, and L. J. Dodson-Dreibelbis. 2003. The external aerodynamics of canine olfaction. In F. G. Barth, J. A. C. Humphrey, and T. W. Secomb, eds. *Sensors and sensing in biology and engineering* (pp. 323–355). New York: SpringerWein.

on the anatomy and sensitivity of the nose:

Harrington, F. H., and C. S. Asa. 2003. Wolf communication. In D. Mech, and L. Boitani, eds. *Wolves: Behavior, ecology and conservation* (pp. 66–103). Chicago: University of Chicago Press.
Lindsay, 2000.
Serpell, 1995.
Wright, R. H. 1982. *The sense of smell.* Boca Raton, FL: CRC Press.

on the vomeronasal organ:

Adams, D. R., and M. D. Wiekamp. 1984. The canine vomeronasal organ. *Journal of Anatomy, 138,* 771–787.

Sommerville, B. A., and D. M. Broom. 1998. Olfactory awareness. *Applied Animal Behavior Science, 57,* 269–286.

Watson, L. 2000. *Jacobson's organ and the remarkable nature of smell.* New York: W. W. Norton & Company.

on human pheromone detection:

Jacob, S., and M. K. McClintock. 2000. Psychological state and mood effects of steroidal chemosignals in women and men. *Hormones and Behavior, 37,* 57–78.

McClintock, M. K. 1971. Menstrual synchrony and suppression. *Nature, 229,* 244–245.

on moist noses:

Mason, R. T., M. P. LeMaster, and D. Muller-Schwarze. 2005. *Chemical signals in vertebrates,* Volume 10. New York: Springer.

on smelling us:

Lindsay, 2000.

on distinguishing twins by scent:

Hepper, P. G. 1988. The discrimination of human odor by the dog. *Perception, 17,* 549–554.

on bloodhounds:

Lindsay, 2000.
Sommerville and Broom, 1998.
Watson, 2000.

on using footsteps to detect trail:

Hepper, P. G, and D. L. Wells. 2005. How many footsteps do dogs need to determine the direction of an odour trail? *Chemical Senses, 30,* 291–298.

Syrotuck, W. G. 1972. *Scent and the scenting dog.* Mechanicsburg, PA: Barkleigh Productions.

on the smell of tuberculosis:
Wright, 1982.

on the smell of disease:
Drobnick, 2006.
Syrotuck, 1972.

on cancer detection:
a partial list of the many studies:
McCulloch, M., T. Jezierski, M. Broffman, A. Hubbard, K. Turner, and T. Janecki. 2006. Diagnostic accuracy of canine scent detection in early- and late-stage lung and breast cancers. *Integrative Cancer Therapies, 5,* 30–39.
Williams, H., and A. Pembroke. 1989. Sniffer dogs in the melanoma clinic? *Lancet, 1,* 734.
Willis, C. M., S. M. Church, C. M. Guest, W. A. Cook, N. McCarthy, A. J. Bransbury, M. R. T. Church, and J. C. T. Church. 2004. Olfactory detection of bladder cancer by dogs: Proof of principle study. *British Medical Journal, 329,* 712–716.

on epileptic seizure detection:
Dalziel, D. J., B. M. Uthman, S. P. McGorray, and R. L. Reep. 2003. Seizure-alert dogs: A review and preliminary study. *Seizure, 12,* 115–120.
Doherty, M. J., and A. M. Haltiner. 2007. Wag the dog: Skepticism on seizure alert canines. *Neurology, 68,* 309.
Kirton, A., E. Wirrell, J. Zhang, and L. Hamiwka. 2004. Seizure-alerting and -response behaviors in dogs living with epileptic children. *Neurology, 62,* 2303–2305.

on urine marking:
Lindsay, 2005.
Lorenz, K. 1954. *Man meets dog.* London: Methuen.

on bladders' single use:

Sapolsky, R. M. 2004. *Why zebras don't get ulcers.* New York: Henry Holt & Company.

on anal sacs:

Harrington and Asa, 2003.
Natynczuk, S., J. W. S. Bradshaw, and D. W. Macdonald. 1989. Chemical constituents of the anal sacs of domestic dogs. *Biochemical Systematics and Ecology, 17,* 83–87.

on anal sacs and vets:

McGreevy, P. (personal communication).

on scratching the ground after marking:

Bekoff, M. 1979. Ground scratching by male domestic dogs: A composite signal. *Journal of Mammalogy, 60,* 847–848.

on antibiotics and smell:

Attributed to John Bradshaw by Coghlan, A. September 23, 2006. Animal welfare: See things from their perspective. NewScientist.com.

on the Manhattan grid:

Margolies, E. 2006. Vagueness gridlocked: A map of the smells of New York. In J. Drobnick, ed., *The smell culture reader* (pp. 107–117). New York: Berg.

on brambish and brunky:

These words were coined by *Calvin and Hobbes* artist Bill Watterson, and put in the mouth of his cartoon tiger Hobbes.

"brilliant smell of water . . .":

Chesterton, G. K. 2004. "The song of the quoodle," in *The collected works of*

G. K. *Chesterton*. San Francisco: Ignatius Press, p. 556. (In this same poem he commented on the relative "noselessness" of man.)

MUTE

"blank bewilderment":

Woolf, V. 1933. *Flush: A biography.* New York: Harcourt Brace Jovanovich, p. 44.

"uncommunicating muteness":

Lamb, C. 1915. *Essays of Elia.* London: J. M. Dent & Sons, Ltd., p. 53.

on the dog's hearing range:

Harrington and Asa, 2003.

on the "Mosquito" teenager repellent:

Vitello, P. June 12, 2006. "A ring tone meant to fall on deaf ears." *The New York Times.*

on alarm clocks:

Bodanis, D. 1986. *The secret house: 24 hours in the strange and unexpected world in which we spend our nights and days.* New York: Simon & Schuster.

on growls:

Faragó, T., F. Range, Zs. Virányi, and P. Pongrácz. 2008. The bone is mine! Context-specific vocalisation in dogs. Paper presented at Canine Science Forum, Budapest, Hungary.

on dog and wolf sounds:

Fox, 1971.
Harrington and Asa, 2003.

on laughs:

Simonet, O., M. Murphy, and A. Lance. 2001. Laughing dog: Vocalizations of domestic dogs during play encounters. Animal Behavior Society conference, Corvallis, OR.

on distinguishing high-pitched sounds:

McConnell, P. B. 1990. Acoustic structure and receiver response in domestic dogs, *Canis familiaris. Animal Behaviour, 39,* 897–904.

on Rico and other vocabularians:

Kaminski, J. 2008. Dogs' understanding of human forms of communication. Paper presented at the Canine Science Forum, Budapest, Hungary.

Kaminski, J., J. Call, and J. Fischer. 2004. Word learning in a domestic dog: Evidence for "fast mapping." *Science, 304,* 1682–1683.

on conversational maxims:

Grice, P. 1975. Logic and conversation. In P. Cole and J. L. Morgan, eds., *Speech acts* (pp. 41–58). New York: Academic Press.

on whimpers, barks, and other vocalizations:

Bradshaw, J. W. S., and H. M. R. Nott. 1995. Social and communication behaviour of companion dogs. In J. Serpell, ed., *The domestic dog: Its evolution, behaviour, and interactions with people* (pp. 115–130). Cambridge: Cambridge University Press.

Cohen, J. A., and M. W. Fox. 1976. Vocalizations in wild canids and possible effects of domestication. *Behavioural Processes, 1,* 77–92.

Harrington and Asa, 2003.

Tembrock, G. 1976. Canid vocalizations. *Behavioural Processes, 1,* 57–75.

on characteristics and kinds of barks:

Molnár, C., P. Pongrácz, A. Dóka, and Á. Miklósi. 2006. Can humans discriminate between dogs on the base of the acoustic parameters of barks? *Behavioural Processes, 73,* 76–83.

Yin, S., and B. McCowan. 2004. Barking in domestic dogs: Context specificity and individual identification. *Animal Behaviour, 68,* 343–355.

on dog bark decibels:

Moffat et al. 2003. Effectiveness and comparison of citronella and scentless spray bark collars for the control of barking in a veterinary hospital setting. *Journal of the American Animal Hospital Association, 39,* 343–348.

"But man himself cannot express love and humility . . .":

Darwin, C. 1872/1965, p. 10.

on hackles:

Harrington and Asa, 2003.

on antithesis:

Darwin, 1872/1965.

on tails:

Bradshaw and Nott, 1995.
Harrington and Asa, 2003.
Schenkel, R. 1947. Expression studies of wolves. *Behaviour, 1,* 81–129.

on posture:

Fox, 1971.
Goodwin, D., J. W. S. Bradshaw, and S. M. Wickens. 1997. Paedomorphosis affects agonistic visual signals of domestic dogs. *Animal Behaviour, 53,* 297–304.

on intentional communication:

Kaminski, J. 2008.

more on urine marking:

Bekoff, M. 1979. Scent-marking by free ranging domestic dogs. Olfactory and visual components. *Biology of Behaviour, 4,* 123–139.
Bradshaw and Nott, 1995.
Pal, S. K. 2003. Urine marking by free-ranging dogs (*Canis familiaris*) in relation to sex, season, place and posture. *Applied Animal Behaviour Science, 80,* 45–59.

DOG-EYED

on the visual range of canids:

Harrington and Asa, 2003.
Miklósi, 2007.

on distribution of photoreceptors in retinae:

McGreevy, P., T. D. Grassia, and A. M. Harmanb. 2004. A strong correlation exists between the distribution of retinal ganglion cells and nose length in the dog. *Brain, Behavior and Evolution, 63,* 13–22.
Neitz, J., T. Geist, and G. H. Jacobs. 1989. Color vision in the dog. *Visual Neuroscience, 3,* 119–25.

on arctic wolves:

Packard, J. 2008. Man meets wolf: Ethological perspectives. Paper presented at Canine Science Forum, Budapest, Hungary.

on Frisbee-catching:

Shaffer, D. M., S. M. Krauchunas, M. Eddy, and M. K. McBeath. 2004. How dogs navigate to catch frisbees. *Psychological Science, 15,* 437–441.

on dogs' recognition of their owners' faces:

Adachi, I., H. Kuwahata, and K. Fujita. 2007. Dogs recall their owner's face upon hearing the owner's voice. *Animal Cognition, 10,* 17–21.

on cows noticing visual details:

Grandin, T., and C. Johnson. 2006. *Animals in translation: Using the mysteries of autism to decode animal behavior.* Orlando, FL: Harcourt.

SEEN BY A DOG

on imprinting in geese:

Lorenz, K. 1981. *The foundations of ethology.* New York: Springer-Verlag.

on newborn and infant humans' visual abilities and development:

The information about infants' visual abilities comes from a century of research. A nice summary is given in Smith, P. K., H. Cowie, and M. Blades. 2003. *Understanding children's development.* Malden, MA: Blackwell Publishing.

on infant tongue protrusion:

Meltzoff, A. N., and M. K. Moore. 1977. Imitation of facial and manual gestures by human neonates. *Science, 198,* 75–78. (They not only stuck out their tongues at day- or even hour-old infants. They also pursed their lips and opened their mouths wide as if in surprise. Even newborns repeated these expressions back at them—or tried to: lip-pursing is probably not a motor ability voluntarily available to the newly born.)

on Kanzi:

Savage-Rumbaugh, S., and R. Lewin. 1996. *Kanzi: The ape at the brink of the human mind.* New York: John Wiley & Sons.

on Alex:

Pepperberg, I. M. 1999. *The Alex studies: Cognitive and communicative abilities of grey parrots.* Cambridge, MA: Harvard University Press.

on the dog-keyboard:

Rossi, A., and C. Ades. 2008. A dog at the keyboard: Using arbitrary signs to communicate requests. *Animal Cognition, 11,* 329–338.

on gaze avoidance:

Bradshaw and Nott, 1995.

on dogs looking at faces:

Miklósi et al., 2003.

on breeders preferring dark eyes:

Serpell, 1996.

on gull fixed action pattern:

Tinbergen, N. 1953. *The herring-gull's world.* London: Collins.

on gaze in human conversation:

Argyle, M., and J. Dean. 1965. Eye contact, distance and affiliation. *Sociometry, 28,* 289–304.
Vertegaal, R., R. Slagter, G. C. Van der Veer, and A. Nijholt. 2001. Eye gaze patterns in conversations: There is more to conversational agents than meets the eyes. In *Proceedings of ACM CHI 2001 Conference on Human Factors in Computing Systems,* Seattle, WA.

on following a pointing gesture:

Soproni, K., Á. Miklósi, J. Topál, and V. Csányi. 2002. Dogs' responsiveness to human pointing gestures. *Journal of Comparative Psychology, 116,* 27–34.

on gaze-following:

Agnetta, B., B. Hare, and M. Tomasello. 2000. Cues to food location that domestic dogs (*Canis familiaris*) of different ages do and do not use. *Animal Cognition, 3,* 107–112.

NOTES AND SOURCES

on attention-getting:

Horowitz, A. 2009. Attention to attention in domestic dog (*Canis familiaris*) dyadic play. *Animal Cognition, 12,* 107–118.

on sonorous mouth-licking:

Gaunet, F. 2008. How do guide dogs of blind owners and pet dogs of sighted owners (*Canis familiaris*) ask their owners for food? *Animal Cognition, 11,* 475–483.

on showing:

Hare, B., J. Call, and M. Tomasello. 1998. Communication of food location between human and dog (*Canis familiaris*). *Evolution of Communication, 2,* 137–159.

Miklósi, Á., R. Polgardi, J. Topál, and V. Csányi. 2000. Intentional behaviour in dog-human communication: An experimental analysis of "showing" behaviour in the dog. *Animal Cognition, 3,* 159–166.

on retrieving games:

Gácsi, M., Á. Miklósi, O. Varga, J. Topál, and V. Csányi. 2004. Are readers of our face readers of our minds? Dogs (*Canis familiaris*) show situation-dependent recognition of human's attention. *Animal Cognition, 7,* 144–153.

on manipulating attention:

Call, J., J. Brauer, J. Kaminski, and M. Tomasello. 2003. Domestic dogs (*Canis familiaris*) are sensitive to the attentional state of humans. *Journal of Comparative Psychology, 117,* 257–263.

Schwab, C., and L. Huber. 2006. Obey or not obey? Dogs (*Canis familiaris*) behave differently in response to attentional states of their owners. *Journal of Comparative Psychology, 120,* 169–175.

on begging experiments:

Cooper, J. J., C. Ashton, S. Bishop, R. West, D. S. Mills, and R. J. Young. 2003. Clever hounds: Social cognition in the domestic dog (*Canis familiaris*). *Applied Animal Behaviour Science, 81,* 229–244.

on attending to a video projection:

Pongrácz, P., Á. Miklósi, A. Doka, and V. Csányi. 2003. Successful application of video-projected human images for signalling to dogs. *Ethology, 109,* 809–821.

on why commands relayed by speakers don't work:

Virányi, Zs., J. Topál, M. Gácsi, Á. Miklósi, and V. Csányi. 2004. Dogs can recognize the behavioural cues of the attentional focus in humans. *Behavioural Processes, 66,* 161–172.

CANINE ANTHROPOLOGISTS

"I am I . . .":

Stein, G. 1937. *Everybody's Autobiography.* New York: Random House, p. 64.

on autistic people using dogs to read others:

Sacks, O. 1995. *An anthropologist on Mars.* New York: Knopf.

on Clever Hans:

Sebeok, T. A., and R. Rosenthal, eds. 1981. *The Clever Hans phenomenon: Communication with horses, whales, apes, and people.* New York: New York Academy of Sciences.

on dogs reading trainers' body movements:

Wright, 1982.

on dogs anticipating us on walks:

Kubinyi, E., Á. Miklósi, J. Topál, and V. Csányi. 2003. Social mimetic behaviour and social anticipation in dogs: Preliminary results. *Animal Cognition, 6,* 57–63.

on distinguishing threatening and friendly strangers:

Vas, J., J. Topál, M. Gácsi, Á. Miklósi, and V. Csányi. 2005. A friend or an enemy? Dogs' reaction to an unfamiliar person showing behavioural cues of threat and friendliness at different times. *Applied Animal Behaviour Science, 94,* 99–115.

NOBLE MIND

on neophilia:

Kaulfuss, P., and D. S. Mills. 2008. Neophilia in domestic dogs (*Canis familiaris*) and its implication for studies of dog cognition. *Animal Cognition, 11,* 553–556.

on physical cognition:

Miklósi, 2007.

on string-pulling:

Osthaus, B., S. E. G. Lea, and A. M. Slater. 2005. Dogs (*Canis lupus familiaris*) fail to show understanding of means-end connections in a string-pulling task. *Animal Cognition, 8,* 37–47.

on use of social cues:

Erdohegyi, A., J. Topál, Zs. Virányi, and Á. Miklósi. 2007. Dog-logic: Inferential reasoning in a two-way choice task and its restricted use. *Animal Behavior, 74,* 725–737.

on dogs looking to humans to solve task:

Miklósi et al., 2003.

on milk-bottle tits:

Fisher, J., and R. A. Hinde. 1949. The opening of milk bottles by birds. *British Birds, 42,* 347–357.

on chickadees experiment:

Sherry, D. F., and B. G. Galef Jr. 1990. Social learning without imitation: More about milk bottle opening by birds. *Animal Behaviour, 40,* 987–989.

on detour learning:

Pongrácz, P., Á. Miklósi, K. Timar-Geng, and V. Csányi. 2004. Verbal attention getting as a key factor in social learning between dog (*Canis familiaris*) and human. *Journal of Comparative Psychology, 118,* 375–383.

on infant imitation:

Gergely, G., H. Bekkering, and I. Király. 2002. Rational imitation in preverbal infants. *Nature, 415,* 755.

Whiten, A., D. M. Custance, J-C. Gomez, P. Teixidor, and K. A. Bard. 1996. Imitative learning of artificial fruit processing in children (*Homo sapiens*) and chimpanzees (*Pan troglodytes*). *Journal of Comparative Psychology, 110,* 3–14.

on dog imitation:

Range, F., Zs. Virányi, and L. Huber. 2007. Selective imitation in domestic dogs. *Current Biology, 17,* 868–872.

on "do it" task:

Topál, J., R. W. Byrne, Á. Miklósi, and V. Csányi. 2006. Reproducing human actions and action sequences: "Do as I Do!" in a dog. *Animal Cognition, 9,* 355–367.

on theory of mind:

Premack, D., and G. Woodruff. 1978. Does a chimpanzee have a theory of mind? *Behavioral and Brain Sciences, 1,* 515–526.

on false belief test:

Wimmer, H., and J. Perner. 1983. Beliefs about beliefs: Representation and constraining function of wrong beliefs in young children's understanding of deception. *Cognition, 13,* 103 128.

on Philip, the dog who informs about the keys:

Topál, J., A. Erdőhegyi, R. Mányik, and Á. Miklósi. 2006. Mindreading in a dog: An adaptation of a primate "mental attribution" study. *International Journal of Psychology and Psychological Therapy, 6,* 365–379.

on the function of play:

Bekoff, M., and J. Byers, eds. 1998. *Animal play: Evolutionary, comparative, and ecological perspectives.* Cambridge: Cambridge University Press.
Fagen, R. 1981. *Animal play behavior.* Oxford: Oxford University Press.

on play-fighting not improving later fighting skills:

Martin, P., and T. M. Caro. 1985. On the functions of play and its role in behavioral development. *Advances in the Study of Behavior, 15,* 59–103.

more on dogs' use of attention, attention-getters, and communication in play:

Horowitz, 2009.

on play signals:

Bekoff, M. 1972. The development of social interaction, play, and metacommunication in mammals: An ethological perspective. *Quarterly Review of Biology, 47,* 412–434.
Bekoff, M. 1995. Play signals as punctuation: The structure of social play in canids. *Behaviour, 132,* 419–429.
Horowitz, 2009.

on the (un)fairness experiment:

Range, F., L. Horn, Zs. Virányi, and L. Huber. 2009. The absence of reward induces inequity aversion in dogs. *Proceedings of the National Academy of Sciences, 106,* 340–345.

INSIDE OF A DOG

on counting:

West, R. E., and R. J. Young. 2002. Do domestic dogs show any evidence of being able to count? *Animal Cognition, 5,* 183–186.

on disjunctive syllogisms:

This is the stoic philosopher Chrysippos of Soloi, according to Bringmann, W., and J. Abresch. 1997. Clever Hans: Fact or fiction? In W. G. Bringmann et al., eds., *A pictorial history of psychology* (pp. 77–82). Chicago: Quintessence.

one of the original scientific attempts to try
to operationalize anthropomorphisms:

Hebb, D. O. 1946. Emotion in man and animal: An analysis of the intuitive process of recognition. *Psychological Review, 53,* 88–106.

on the suprachiasmatic nucleus:

A nice review of some recent work: Herzog, E. D., and L. J. Muglia. 2006. You are when you eat. *Nature Neuroscience, 9,* 300–302.

on changes in sleep with age:

Takeuchi, T., and E. Harada. 2002. Age-related changes in sleep-wake rhythm in dog. *Behavioural Brain Research, 136,* 193–199.

on the movement of smells in a room:

Bodanis, 1986.
Wright, 1982.

on bees' sense of time:

Boisvert, M. J., and D. F. Sherry. 2006. Interval timing by an invertebrate, the bumble bee *Bombus impatiens. Current Biology, 16,* 1636–1640.

"boredom is rarely discussed in the non-human scientific literature":

But see Wemelsfelder, F. 2005. Animal Boredom: Understanding the tedium of confined lives. In F. D. McMillan, ed., *Mental health and well-being in animals* (pp. 79–91). Ames, Iowa: Blackwell Publishing.

"man is the only animal who can be bored":

Fromm, E. 1947. *Man for himself, an inquiry into the psychology of ethics.* New York: Rinehart, p. 40.

on the mirror test:

Gallup, G. G. Jr. 1970. Chimpanzees: Self-recognition. *Science, 167,* 86–87.
Plotnik, J. M., F. B. M. de Waal, and D. Reiss. 2006. Self-recognition in an Asian elephant. *Proceedings of the National Academy of Science, 103,* 17053–17057.
Reiss, D., and L. Marino. 2001. Mirror self-recognition in the bottlenose dolphin: A case of cognitive convergence. *Proceedings of the National Academy of Science, 98,* 5937–3942.

on sheepdogs' knowing they are not sheep:

Coppinger and Coppinger, 2001.

Snoopy quote:

Gesner, C. 1967. *You're a good man, Charlie Brown: Based on the comic strip* Peanuts *by Charles M. Schulz.* New York: Random House.

on scrub-jay caching:

Raby, C. R., D. M. Alexis, A. Dickinson, and N. S. Clayton. 2007. Planning for the future by western scrub-jays. *Nature, 445,* 919–921.

on ontogenetic ritualization:

Tomasello, M., and J. Call. 1997. *Primate cognition.* New York: Oxford University Press.

on medieval punishment of dogs:

Evans, E. P. 1906/2000. *The criminal prosecution and capital punishment of animals.* Union, NJ: Lawbook Exchange, Ltd.

on owners thinking dogs know right and wrong:

Pongrácz, P., Á. Miklósi, and V. Csányi. 2001. Owners' beliefs on the ability of their pet dogs to understand human verbal communication: A case of social understanding. *Cahiers de psychologie, 20,* 87–107.

on teddy-bear guard-dog:

Kennedy, M. August 3, 2006. "Guard dog mauls Elvis's teddy in rampage." *The Guardian.*

on guilt experiments:

Horowitz, A. 2009. Disambiguating the "guilty look": Salient prompts to a familiar dog behaviour. *Behavioural Processes, 81,* 447–452.
Vollmer, P. J. 1977. Do mischievous dogs reveal their "guilt"? *Veterinary Medicine, Small Animal Clinician, 72,* 1002–1005.

on the blind Labrador Norman:

Goodall, J., and M. Bekoff. 2002. *The ten trusts: What we must do to care for the animals we love.* New York: HarperCollins.

on emergency experiment:

Macpherson, K., and W. A. Roberts. 2006. Do dogs (*Canis familiaris*) seek help in an emergency? *Journal of Comparative Psychology, 120,* 113–119.

"What is it like to be a bat?":

Nagel, T. 1974. What is it like to be a bat? *Philosophical Review, 83,* 435–450.

NOTES AND SOURCES

on Stanley's view of the world:
Sterbak, J. 2003. "From here to there."

on personal space:
Argyle and Dean, 1965.

on differences in heeling styles:
Packard, 2008.

on a snail's perception of a tapping stick:
von Uexküll, 1957/1934.

on pressure release as reinforcement in horses:
McGreevy and Boakes, 2007.

on slaughterhouse design:
Grandin and Johnson, 2005.

on perception of objects under yellow light:
I owe my understanding of the blood-draining effect of yellow light to the exhibit "Room for one colour" by the artist Olafur Eliasson, in which he lights a room by bulbs emitting an extremely narrow range of what appears as yellow light.

Wittgenstein on dogs:
Wittgenstein, L. 1953. *Philosophical investigations.* New York: Macmillan.

on the length of a moment:
von Uexküll, 1957/1934.

on clicker training:

McGreevy and Boakes, 2007.

on the wolves' provocative showing of food:

Miklósi, 2007.

YOU HAD ME AT HELLO

on vasopressin in the prairie vole:

Alcock, J. 2005. *Animal behavior: An evolutionary approach,* 8th ed. Sunderland, MA: Sinauer Associates.

on sheepdog imprinting:

Coppinger and Coppinger, 2001.

on not all animals being equally anthropomorphizable:

Eddy, T. J., G. G. Gallup Jr., and D. J. Povinelli. 1993. Attribution of cognitive states to animals: Anthropomorphism in comparative perspective. *Journal of Social Issues, 49,* 87–101.

on our attraction to infants and other neotonized creatures:

Gould, S. J. 1979. Mickey Mouse meets Konrad Lorenz. *Natural History, 88,* 30–36.

Lorenz, K. 1950/1971. Ganzheit und Teil in der tierischen und menschlichen Gemeinschaft. Reprinted in R. Martin, ed., *Studies in animal and human behaviour,* vol. 2 (pp. 115–195). Cambridge, MA: Harvard University Press.

"we need the eggs":

Said by Woody Allen's alter ego Alvy Singer in *Annie Hall,* 1977.

on biophilia:

Wilson, E. O. 1984. *Biophilia*. Cambridge, MA: Harvard University Press.

on touch:

Lindsay, 2000.

on Harlow studies:

Harlow, H. F. 1958. The nature of love. *American Psychologist, 13,* 673–685.
Harlow, H. F., and S. J. Suomii. 1971. Social recovery by isolation-reared monkeys. *Proceedings of the National Academy of Sciences, 68,* 1534–1538.

on the alleviation of puppies' distress with soft toys:

Elliot, O., and J. P. Scott. 1961. The development of emotional distress reactions to separation in puppies. *Journal of Genetic Psychology, 99,* 3–22.
Pettijohn, T. F., T. W. Wong, P. D. Ebert, and J. P. Scott. 1977. Alleviation of separation distress in 3 breeds of young dogs. *Developmental Psychobiology, 10,* 373–381.

"thermotactile sensory probe":

Fox, M. 1971. Socio-infantile and socio-sexual signals in canids: A comparative and developmental study. *Zeitschrift fuer Tierpsychologie, 28,* 185–210.

on our tactile resolution:

Attributed to the psychophysicist Ernst Heinrich Weber by von Uexküll (1957/1934).

on whiskers:

Lindsay, 2000.

"redirected appeasement ceremony":

Lorenz, K. 1966. *On aggression.* New York: Harcourt, Brace & World, Inc., p. 170.

on guide dogs and the blind:

Naderi, Sz., Á. Miklósi, A. Dóka, and V. Csányi. 2001. Cooperative interactions between blind persons and their dog. *Applied Animal Behavior Sciences, 74,* 59–80.

on dog-human play:

Horowitz, A. C., and M. Bekoff. 2007. Naturalizing anthropomorphism: Behavioral prompts to our humanizing of animals. *Anthrozoös, 20,* 23–35.

on timing patterns of flirters:

Sakaguchi, K., G. K. Jonsson, and T. Hasegawa. 2005. Initial interpersonal attraction between mixed-sex dyad and movement synchrony. In L. Anolli, S. Duncan Jr., M. S. Magnusson, and G. Riva, eds., *The hidden structure of interaction: From neurons to culture patterns* (pp. 107–120). Amsterdam: IOS Press.

on synchrony between dogs and people:

Kerepesi, A., G. K. Jonsson, Á. Miklósi, V. Csányi, and M. S. Magnusson. 2005. Detection of temporal patterns in dog–human interaction. *Behavioural Processes, 70,* 69–79.

on dogs' sensitivity to cortisol and testosterone:

Jones, A. C., and R. A. Josephs. 2006. Interspecies hormonal interactions between man and the domestic dog (*Canis familiaris*). *Hormones and Behavior, 50,* 393–400.

on dogs' sensitivity to play styles:

Horváth, Zs., A. Dóka, and Á. Miklósi. 2008. Affiliative and disciplinary behavior of human handlers during play with their dog affects cortisol concentrations in opposite directions. *Hormones and Behavior, 54,* 107–114.

on lowered blood pressure, other measures, and hormone changes:

Friedmann, E. 1995. The role of pets in enhancing human well-being: Physiological effects. In I. Robinson, ed., *The Waltham book of human-animal interactions: Benefits and responsibilities of pet ownership* (pp. 35–59). Oxford: Pergamon.

Odendaal, J. S. J. 2000. Animal assisted therapy—magic or medicine? *Journal of Psychosomatic Research, 49,* 275–280.

Wilson, C. C. 1991. The pet as an anxiolytic intervention. *Journal of Nervous and Mental Disease, 179,* 482–489.

on other dog-owning benefits:

Serpell, 1996.

on contagious yawns:

Joly-Mascheroni, R. M., A. Senju, and A. J. Shepherd. 2008. Dogs catch human yawns. *Biology Letters, 4,* 446–448.

on Derrida, naked, and his cat:

Derrida, J. 2002. L'animal que donc je suis (à suivre). Translated as "The animal that therefore I am (more to follow)." *Critical Inquiry, 28,* 369–418.

THE IMPORTANCE OF MORNINGS

on herding and the "eye":

Coppinger and Coppinger, 2001.

on handedness in dogs:

P. McGreevy, personal communication.

on training:

See McGreevy and Boakes, 2007, for some ideas.

on preference for the new:

Kaulfuss and Mills, 2008.

on dog gaits:

Brown, 1986.

"colors one's thinking about it forever afterwards":

So said George Schaller, whose many books are full of named animals. Quoted in Lehner, P. 1996. *Handbook of ethological methods,* 2nd ed. Cambridge: Cambridge University Press, p. 231.

on zebra-finch leg-band preferences:

Burley, N. 1988. Wild zebra finches have band colour preferences. *Animal Behaviour, 36,* 1235–1237.

Acknowledgments

Of the following dogs:

No one who knew Pumpernickel will be surprised that my most ardent thanks go to her, for choosing us at the shelter and for allowing me the incredible pleasure of knowing her. I have thanked her many times since, with cheese taking over where words failed me. Thanks to Finnegan, for being his own dog, and for being such an utterly doggish dog. Every day is improved to have him come running madly toward me. Thanks to the dogs of yore: to Aster, who endured a lot of childhood foolishness and taught me how to be less foolish; to Chester, who could grin and growl at the same time; to Beckett and Heidi, who in death highlighted what is precious; and to Barnaby, who in catness highlighted what is dog.

Of the following people:

One hears that books are difficult to write. If so, this is not a book, for it was a delight to write, as it is delightful to observe and be with dogs and think dog-thoughts full-time. I was even more pleased that I would be handing the book over to the people at

333

Scribner, whom I could count on to make my bagful of chapters into an actual book. I am indebted to Colin Harrison for his tireless reading of drafts and for being open to just about anything. Had I turned a book about dogs into one about cats I suspect Colin would have acceded to it . . . as long as it was still a good read. Many thanks to Susan Moldow for her enthusiasm from the very beginning.

Before I had an agent I scanned acknowledgments pages for those that included words that would send me scurrying to shine up a proposal to their agents. Sorry, Kris Dahl, in advance, for this: she is the very person you want representing you and your book; and I thank her.

My graduate school advisors and mentors, Shirley Strum and Jeff Elman, were willing to consider how an abstruse theoretical question about cognition could be addressed by observations of dogs—and they improved the theory and the practice. I was and am still appreciative. Thanks to Aaron Cicourel, who is also, as he says, one of those folks who try to saw through wood the hard way. Marc Bekoff was one of the first to treat dog play as biologically interesting. It was his writing (with the very keen Colin Allen), and later his advice, devotion, and friendship that led me to pursue my own research.

I owe thanks to Damon Horowitz, with whom I hatched the plan to write this book, and who seemed to believe that it was a sage and realistic idea. His consummate skepticism about all matters is balanced by his unfettered support of all that matters to me. I owe pretty much everything to my parents, Elizabeth and Jay. They were the first people I wanted to show the book to, for all the right reasons. As for you, Ammon Shea: you make me better with words, you make me better with dogs, and you make me better.

Index

Turn the page for a Q & A
with author Alexandra Horowitz

An interview with Alexandra Horowitz, author of *Inside of a Dog*

What drove you to write this book?

I spent many years living with my own dog, Pumpernickel, and had a range of questions about her behavior and experience that will be familiar to anyone who has ever shared space with a dog: What does she do when I'm away from home? Is she bored? Happy? What does she dream about? Why does she roll in *that*? Pump was a great, unique character, and the longer I knew her, the more curious I became about her.

At the same time, I was working toward my doctorate in cognitive science. I became interested in what is now called "animal cognition": observing the behavior of animals to get an idea of their cognitive capacities. Notably, dogs were not a subject of study when I began: only apes and monkeys (and some exceptional birds and so on) were considered to be cognitively interesting—presumably because of their close relation to humans. But it occurred to me, and to a few other scientists at the same time, that dogs could be studied in the same way. That's when I began doing research observing dogs.

Since then, the study of dog cognition has taken off: there are now dozens of academic groups looking at dog behavior. Still, most academic research doesn't try to answer the kinds of questions I had about my own dog. I wrote the book as a way to make the recent research accessible to those interested in dogs, and to try to apply it toward those questions.

How is your book different than other dog books? Does the world need another book on dogs?

Despite the fabulous photo of a Great Dane's head on the cover (and back) of the book jacket, I really don't think of my book as a typical "dog book." It is a book about using cognitive science to better imagine the minds of animals—and the animal I focus on is the dog. It is also an attempt to answer the question "What is it like to be another animal?"—a philosopher's question, but one that I think many people have about their pets or other animals they run across.

Within the category of books on dogs, I think there is a lot of territory that hasn't been covered yet. We are awash in training books, and in personal stories of cute, or bad, or heroic, or clever dogs. My book is not one of these. It is not a training book (though through a better understanding of her dog, an owner may come to train him better); and it is not a sentimental book (though it is full of the sentiment that comes with having a relationship with one of these magnificent creatures). Instead, this book is about imagining the dog's point of view: how the dog experiences the world; what he wants and needs; what he thinks about and understands. I think this is something we haven't done nearly enough of, especially considering how prevalent dogs are in our society, and in our days.

Could you explain the concept of a dog's *umwelt*, which is a centerpiece of your book?

An animal's *umwelt* is what life is like to the animal: the animal's point of view. The idea is that to understand an animal, one has to appreciate how the world looks to the animal. And to do that, one needs to know what sensory equipment the animal has—How good is his vision? What can he smell? Can he detect electrical impulses? and so on—and the things in the world that are important to them. Humans are a big part of the *umwelten* of dogs—but in a housefly's umwelt, for instance, we are pretty much indistinguishable from other mammals. On the other hand, the dog and the fly both share an acute perception of, and a fascination in locating, foul-smelling objects—whereas such smells register to us, but only with a mind to avoid the smell's source.

In my book I encourage the reader to try to understand the dog better by paying more attention to his umwelt. What can the dog see, smell, hear? What does the dog think about and know about? What things are relevant to the dog, and what things are not? To grasp the dog's umwelt is to better appreciate what it is like to be a dog.

Can we know what is it like to be a dog, then? Do they see the world like we do?

I think it is not possible to know *exactly* what it's like to be a dog, just as it is impossible to know what it is like to be another person. But the more we know about the dog's abilities, both cognitively and perceptually, the better we are able to imagine what it might be like to be a dog.

We naturally imagine that dogs are more or less like us—only less sophisticated, less smart, with less going on in their heads. This is simply wrong. When we realize what they can sense that we cannot, a new picture appears: one in which the dog is in an extraordinarily rich sensory world, with complex social interactions, and with a special ability to read our behavior. Dogs *don't* see the world like we do: they "see" mostly through smell—both through the nose and a special organ called the "vomeronasal organ" in the roof of their mouths. Their vision is pretty good, not as finely detailed and colored as ours is, but it is secondary to their ability to see the world through their noses. Even imagining that is difficult for us vision-centered folks.

You write that we often misinterpret dogs' behavior. Can you give an example of how we do so?

Dogs are frequently treated as though all their behaviors map to human behavior. We call raising a paw "shaking hands"—this is tongue-in-cheek, of course, but it is still surprising to learn that "shaking" is a submissive behavior of dogs, done to show that they are not threatening, and to avoid an attack. I certainly don't think that's what people intend to have the dog say with a shake.

My favorite example, one that is in the book, is of the dog "kiss": a dog's slobbering, rambunctious licking of our mouths when we return from being away is often considered to be a sign of his affection for us. But if we look at the behavior of their cousins and ancestors, wolves, we get a far different impression. When a wolf returns to the pack from a hunt, he or she is mobbed by his packmates—who all lick madly at his mouth. What they are trying to do is to get the returning wolf to regurgitate some of the freshly killed meat he has eaten (which they often do).

So when your dog licks your mouth, he is probably doing

something similar: seeing what you've eaten, and encouraging you to spit some of it up (they will never be unhappy if you do . . . unlike the others in your life who may kiss you on the mouth). On the other hand, it is still fair to call this behavior a "greeting" behavior, one that despite its gory past is also indication of recognition, familiarity, and—perhaps!—affection.

You write "dogs are anthropologists among us." What do they know about us?

Based on smell alone, they seem to know a lot about individuals. They can tell if you've recently had sex, smoked a cigarette, had a bath; they know if you've just eaten, or gone for a run, or petted another dog. They can smell your emotions: dogs have the ability to sense the hormones we exude when we are scared; they can most likely detect other emotions too.

Smell is not their only source of information about us. Dog owners are sometimes impressed with how dogs know when they are packing for a trip, or getting ready for a walk. This is just the tip of the iceberg. Humans are creatures of habit, tending to act similarly when we get dressed, get ready to go out, prepare dinner, and so on. Dogs are very good at observing the series of events that leads to a consequence of interest (like a walk), and remembering the chain of events that preceded it. Sometimes it seems that dogs know our intent before even we do.

What do you hope readers take away from your book?

I hope people gain a new appreciation of just how different dogs are from what we ordinarily think—and that people use this to build a new relationship with their dogs based on what the dog can understand and is interested in. I hope that people start taking their dog's umwelt into account—and thus reconsider putting that raincoat on him, or pulling him away from a good smell, or keeping him from socializing with other dogs.

When people get a dog, one of the first things they set about doing is figuring out how to "train" him. I find this curious—somewhat like schooling a newborn infant in the house rules as soon as he's home from the hospital. There are so many more compelling ways of dealing with dogs than just training them and then considering the interaction complete. If, instead, we live with them for a while, watch them, let them act doggily, and let them react to us and us react to them, we begin to forge a relationship that is far more interesting for all involved.

You've highlighted the importance of attention to the dog's experience. What can owners do to improve the experience of their dogs?

All owners presumably already know that every dog needs a daily dose of attention—and not only in the form of food. Regular exercise and interaction (with humans or dogs) is not just nice, it is essential, as every dog expert will tell you.

Beyond that, there is a lot owners can do to creatively expand their dogs' horizons. I've heard from many readers who liked the idea of the "smell walk." Some of them may have already been slowing down a lot to let their dogs sniff, and now they simply do it with enjoyment, instead of with annoyance, knowing that the dog is amassing a rich collection of olfactory notes to ruminate on.

Most dogs spend a goodly portion of the day asleep, but this is not only because they are physically different than us. It is also the result of the mind-numbing boredom of being left with nothing to do and no one to interact with while we are absent.

When you leave your dog at home, preparing a treasure hunt of small treats hidden around the house is a fine way to amuse and distract him for part of the day.

I have found that creating shared, elaborate rituals is greatly amusing to my pups. The "rituals" are, at core, just normal habits: things we do day after day. I have consciously added extra steps—sometimes even unrelated steps—to make a long string of events that reliably occur one after the other. For instance, lately Finnegan and I have been rising early to have a walk; when we return, my husband and young son are still asleep. I feed Finn, then myself, and we have a protracted session of playing with whatever ripe, filthy ball he's found in the park. Then I put the ball away, have myself a cup of coffee, and look over at Finn. He's on the couch, watching me. All I have to do is rise while looking at him and he knows: we're off to wake the boys. He can hardly contain himself as we open the bedroom door, and he prances around on the bed, waking everyone up.

The satisfaction that he gets from this is evident. Not only are many elements of the sequence engaging; the whole fact of the sequence is interesting. Ever since I noticed how attuned he was to what happened next, I made sure to follow the strict order of the activities. Now Finn not only anticipates the next activity in order, but he gets excited when we *diverge* from the order—by, say, going in to wake them up before breakfast. Either way is greatly entertaining.

Have there been any new dog-cognition studies since you published the book in 2009?

Lots. Some are entirely new paradigms; some go into more specifics about dogs' abilities. One study showed, for instance, that though all dogs are good at understanding a human point, working dogs and dogs with small heads and forward-facing eyes are better at it than other breeds. Another study looking at dogs' attention to us found that dogs look more at one side—the *right* side—of our faces than the other: a bias in looking also found in us, when looking at each other. Currently I am interested in studying whether dogs distinguish between humans who dole out food fairly and those who are unfair (even when the dog gets the same amount himself). In essence, we are asking if dogs have a sense of *justice*.

What is exciting to see is that any question we have about the dog can be formulated into a testable hypothesis. Now, the answer that researchers get is not always the one that owners want to hear: especially those results which indicate that dogs know less than we think they do. But that is the value of science: it supports our intuition in some cases, and defies it in others. In all cases, it is only as good as its method. And we are only talking about "dogs"—some imaginary, "average" dog. It is impossible for science to capture the uniqueness that every owner, including me, knows characterizes her own dog.

About the Author

Alexandra Horowitz earned her B.A. in philosophy from the University of Pennsylvania and a Ph.D. in cognitive science from the University of California at San Diego, studying dog cognition. She is currently a term assistant professor of psychology at Barnard College and continues to research dog behavior. In addition to her work with dogs, she has also studied cognition in humans, rhinoceroses, and bonobos. She previously worked as a lexicographer at Merriam-Webster and a fact-checker for *The New Yorker*. She lives in New York City with her husband, son, and Finnegan, a dog of indeterminate parentage and determinate character.

She also likes to sketch her dogs.